T0202175

Leibniz: General Inquiries on the Analysis of Notions and Truths

BSHP NEW TEXTS IN THE HISTORY OF PHILOSOPHY

The aim of this series is to encourage and facilitate the study of all aspects of the history of philosophy, including the rediscovery of neglected elements and the exploration of new approaches to the subject. Texts are selected on the basis of their philosophical and historical significance and with a view to promoting the understanding of currently under-represented authors, philosophical traditions, and historical periods. They include new editions and translations of important, yet less well-known works which are not widely available to an Anglophone readership. The series is sponsored by the British Society for the History of Philosophy (BSHP) and is managed by an editorial team elected by the society. It reflects the society's main mission and its strong commitment to broadening the canon.

General editors
Maria Rosa Antognazza
Michael Beaney
Mogens Lærke (managing editor)

Leibniz

General Inquiries on the Analysis of Notions and Truths

Edited with an English translation by

MASSIMO MUGNAI

OXFORD
UNIVERSITY PRESS

OXFORD

UNIVERSITY PRESS

Great Clarendon Street, Oxford, OX2 6DP,
United Kingdom

Oxford University Press is a department of the University of Oxford.
It furthers the University's objective of excellence in research, scholarship,
and education by publishing worldwide. Oxford is a registered trade mark of
Oxford University Press in the UK and in certain other countries

Published in the United States of America by Oxford University Press
198 Madison Avenue, New York, NY 10016, United States of America

British Library Cataloguing in Publication Data
Data available

Library of Congress Control Number: 2020952439

ISBN 978-0-19-289590-5

Printed and bound in Great Britain by
CPI Group (UK) Ltd, Croydon, CR0 4YY

Links to third party websites are provided by Oxford in good faith and
for information only. Oxford disclaims any responsibility for the materials
contained in any third party website referenced in this work.

Contents

List of Abbreviations

A	G. W. Leibniz (1923–). *Sämtliche Schriften und Briefe.* Darmstad, Leipzig, and Berlin: Deutsche Akademie der Wissenschaften zu Berlin.
DAC	G. W. Leibniz (2020). *Dissertation on Combinatorial Art*, ed. M. Mugnai, H. van Ruler, and M. Wilson. Oxford: Oxford University Press.
Discourse	G. W. Leibniz (2020). *Discourse on Metaphysics*, ed. Gonzalo Rodriguez-Pereyra. Oxford: Oxford University Press.
GM	G. W. Leibniz (1849–63). *Mathematische Schriften*, ed. C. I. Gerhardt, 7 vols. Berlin: A. Asher and Halle: H. W. Schmidt.
GP	G. W. Leibniz (1875–90). *Die Philosophischen Schriften*, ed. C. I. Gerhardt, 7 vols. Berlin: Weidmannsche Buchhandlung.
Grua	G. W. Leibniz (1948). *Textes inédits d'après les manuscrits de la Bibliothèque provinciale de Hanovre*, ed. G. Grua. 2 vols. Paris: Presses Universitaires de France.
L	G. W. Leibniz (1969). *Philosophical Papers and Letters. A Selection*, tr. and ed. Leroy E. Loemker. 2nd edn. Dordrecht: D. Reidel.
LH	'Leibniz-Handschriften'. Niedersächsische Landesbibliothek, Hanover.
LP	G. W. Leibniz (1966). *Logical Papers. A Selection*, tr. and ed. G. H. R. Parkinson. Oxford: Clarendon Press.
MP	G. W. Leibniz (1973). *Philosophical Writings*, ed. G. H. R. Parkinson, tr. Mary Morris and G. H. R. Parkinson. London: J. M. Dent & Sons.
Mugnai (2008)	G. W. Leibniz (2008). *Ricerche generali sull'analisi delle nozioni e delle verità e altri scritti di logica*, ed. Massimo Mugnai. Pisa: Edizioni della Normale.
NE	G. W. Leibniz (1981). *New Essays on Human Understanding*, tr. and ed. Peter Remnant and Jonathan Bennett. Cambridge: Cambridge University Press.

Opuscules	Louis Couturat, ed. (1903). *Opuscules et fragments inédits de Leibniz, extraits des manuscrits de la Bibliothèque royale de Hanovre.* Paris: Alcan, 1903.
Philosophical Essays	G. W. Leibniz (1989). *Philosophical Essays*, ed. and tr. Roger Ariew and Daniel Garber. Indianapolis, IN, and Cambridge: Hackett Publishing Company.
Rauzy (1998)	G. W. Leibniz (1998). *Recherches générales sur l'analyse des notions et des vérités. 24 thèses métaphysiques et autres textes logiques et métaphysiques*, ed. Jean-Baptiste Rauzy. Paris: Presses Universitaires de France.
Schriften zur Syllogistik	G. W. Leibniz (2019). *Schriften zur Syllogistik*, ed. and tr. Wolfgang Lenzen. Hamburg: Felix Meiner Verlag.
Schupp (1982)	G. W. Leibniz (1982). *Generales Inquisitiones de Analysi Notionum et Veritatum—Allgemeine Untersuchungen über die Analyse der Begriffe und Wahrheiten*, ed. F. Schupp. Hamburg: Meiner Verlag.
T	G. W. Leibniz (1985). *Theodicy. Essays on the Goodness of God, the Freedom of Man and the Origin of Evil*, ed. Austin Farrer, tr. E. M. Huggard. Oxford: Oxford University Press.

On This Edition

The original manuscript of the *GI* was composed employing different inks and pen nibs, clear evidence, together with the marginal notes and the many erasures, that Leibniz revised it several times. Some paragraphs, for instance, are crossed out and completely rewritten on the margin of the sheet. This shows that Leibniz pondered at length on the content of the *GI*, which is his most complete work entirely devoted to the project of building a new logic.

The text of the *GI* was published for the first time in 1903 by Louis Couturat in his collection of Leibniz's essays (*Opuscules*: 356–99). A German translation based on Couturat's edition appeared in 1960 in a collection of Leibniz's texts on logic published by Franz Schmidt: Gottfried Wilhelm Leibniz, *Fragmente zur Logik*, Berlin: Akademie Verlag. Translations into English were made by George H. R. Parkinson: Leibniz, *Logical Papers*, Oxford: Clarendon Press, 1966; and Walter H. O'Briant, Gottfried Wilhelm Leibniz's *General Investigations Concerning the Analysis of Concepts and Truths. A Translation and an Evaluation*, Athens, GA: University of Georgia Press, 1968.

For a first critical edition we had to wait until 1982, when Franz Schupp published the Latin text revised from the original manuscript, together with a German translation of it. The definitive critical text was established by the Berlin Academy in 1999 (G. W. Leibniz, *Sämtliche Schriften und Briefe*, Berlin: Akademie Verlag, VI Reihe, IV, pp. 739–88).

Compared with Couturat's pioneering work, Schupp's text was a considerable improvement (it supplied passages that were omitted in the Couturat edition and corrected several mistakes due to a wrong transcription). The Academy edition, in its turn, improved on Schupp's edition, establishing the correct reading of some words and sentences that were misread by Schupp. In general, however, Schupp's edition continues to be quite reliable and, in some respects, there are even reasons to prefer it to the Academy edition. Leibniz, for instance, is aware of the

distinction between *use* and *mention* applied to a word or a sentence, and, aiming to stress this difference, he usually underlines the items that are mentioned and not used. Unfortunately, however, he does not systematically employ this device, and uses it even with the aim of emphasizing some expressions. In the manuscript, for instance, we find phrases like 'terms <u>coincide</u> if they can be substituted…', where Leibniz underlines the word 'coincide' for emphasis, not aiming to stress that it has been mentioned and not used.[1]

The editors of the Academy edition have adopted the policy of rendering in a different font words underlined by Leibniz and italicize all words that they think are mentioned. The problem is, however, that they make very extensive use of italics and, in some cases, they blur the difference between words used and words mentioned.

Another problem with the Academy edition is that all Leibniz's marginal notes in the manuscript become footnotes in the printed text. Now, some of these notes are clearly remarks that Leibniz wrote when doing a second reading of the text and have, as it were, a 'personal' character, something like 'here I have to add this and this', or 'NB', or 'this paragraph needs to be revised', etc. But there are other remarks which integrate into the text and are written as parts of it proper. The Academy edition, however, does not attempt to discriminate between these two kinds of marginal notes.

The present translation has been conducted on the basis of the original manuscript of the *GI* and of the text established by the Academy of the Sciences of Berlin, taking obvious advantage of the pre-existing translations by Parkinson and O'Briant. In translating the text into English, an attempt has been made to employ the distinction use–mention economically and to distinguish the marginal notes that are supposed to integrate into the text from those that seem to be of a different nature.

---

[1] Cf. *LP*: lxiii–lxiv: "One difficulty which faces the translator of Leibniz concerns the use of quotation marks. These, now commonly employed to indicate that a word or group of words is being mentioned as opposed to being used, are not used at all by Leibniz, who has no standard way of indicating the mention as opposed to the use of a word or words. Sometimes he uses a capital letter.… Sometimes he underlines a word or phrase.… Sometimes he uses parentheses.… Sometimes he uses the Greek definite article, followed by the word or words mentioned."

The Latin text does not include the transcription of passages or words that Leibniz first wrote and then deleted: these are included in the Academy edition.

Key to the symbols:

(1) Words or phrases in square brackets, [...], have been corrected or integrated by the editor.

(2) Words or phrases in angle brackets, ⟨...⟩, have been written in the margin of the manuscript or added to the text by Leibniz.

(3) Words or phrases in both square brackets and angle brackets, [⟨...⟩], correspond to corrections or conjectures proposed by the Academy edition of the *GI*.

(4) 'L' in the footnotes refers to the original word (or words) employed by Leibniz.

Introduction

1 Genesis and Structure of the 'General Inquiries'

Leibniz composed the *General Inquiries on the Analysis of Notions and Truths* [*Generales inquisitiones de analysi notionum et veritatum*] ('GI' henceforth) during the year 1686, the same year in which he began to correspond with Arnauld and wrote the *Discourse on Metaphysics*.[1] The correspondence with the philosopher and theologian Antoine Arnauld (1612–94) constitutes one of the main sources for the study of Leibniz's philosophy, and the same holds for the *Discourse*, which offers a first systematic account of notions like those of *complete concept* of an individual, *pre-established harmony* between soul and body, and *substantial form* (something very similar to the 'monad' of Leibniz's mature philosophy).[2]

The *GI* is a necessary supplement to the correspondence with Arnauld and the *Discourse* in so far as it develops a central topic of Leibniz's metaphysics and shows the intimate connection that links Leibniz's philosophy with the attempt to create a new kind of logic. It is in the *GI*, indeed, that Leibniz articulates for the first time his favourite solution to the problem of contingency, and it is in the *GI* that he displays the main features of his logical calculus.

At first glance, the *GI* gives the impression of a 'compact' and coherent work: it begins with a fairly long introduction where several topics are discussed (philosophy of logic, metaphysics, and grammar), and then a list of paragraphs of various lengths follows, marked with numbers from 1 to 200. To the sequence of paragraphs, however, there is no corresponding systematic and coherent development of a logical calculus. It is only towards the end of the essay that Leibniz proposes a set of principles from which the theorems previously proved can be derived; and he

---

[1] Cf. Antognazza (2009: 239–41).   [2] Cf. *Discourse* and A VI, 4B: 1529–88.

attains this result without explicitly discussing the relationship of his final outcome with the other principles previously proposed: these simply survive in the body of the text as evidence of the steps that have evolved to produce the final outcome.[3]

On at least two occasions he revives his old project of employing numbers to express propositions but then, after a while, he abandons this issue and abruptly begins to develop a different topic.[4] As George Parkinson remarked, the *GI* is a difficult work 'in which Leibniz often seems to be groping his way'.[5] This, however, does not undermine the extraordinary value of the *GI*, which is very rewarding for everyone interested in logic, philosophy of logic, and metaphysics (besides Leibniz's thought). As Marko Malink and Anubav Vasudevan point out, the *GI*:

> does not take the form of a methodical presentation of an antecedently worked-out system of logic, but rather comprises a meandering series of investigations covering a wide range of topics. As a result, it can be difficult to discern the underlying currents of thought that shape the treatise amidst the varying terminology and conceptual frameworks adopted by Leibniz at different stages of its development.[6]

Given this composite structure of the *GI*, in what follows I devote two sections to introduce each of the two main topics of this work: logic and metaphysics.

Section 2 ('Logic') begins with a preparatory account of Leibniz's project for a universal characteristic and focuses on the relationships between *rational grammar* and logic. Then, I will discuss the general structure and the main ingredients of Leibniz's logical calculus as presented in the *GI*.

Section 3 ('Metaphysics') is centred on the problem of contingency, which caused a lot of trouble for Leibniz from the beginning of his correspondence with Arnauld until the end of his life. I attempt to explain,

---

[3] Malink and Vasudevan (2016: 685–6).

[4] Leibniz's idea of employing numbers to designate concepts (and propositions) traces back to the *DAC* (1666): 4–5; 161. An extensive discussion of Leibniz's use of numbers in his logical essays can be found in chapter 3 of *Schriften zur Syllogistik*.

[5] *LP*: xxvi–xxvii.   [6] Malink and Vasudevan (2016: 686).

first, the nature of this problem and then to show how Leibniz reckoned he had solved it: in the *GI*, indeed, we find, even though it is expressed in a tentative way, the core of his solution based on infinite analysis.

2 Logic

2.1 The Characteristic Art and the Rational Grammar

The *GI* is an essential part of the project for the constitution of what Leibniz calls *characteristic art* (*ars characteristica*). In Latin the word *character* means 'sign' or 'mark', and the *characteristic art* was conceived as a system of signs provided with rules for performing three different tasks: encoding concepts, forming propositions, and inferring propositions from propositions.

The first embryonic idea of the characteristic art can be traced back to the *Dissertation on Combinatorial Art*, which Leibniz wrote when he was 19 years old (1665–6).[7] The *Dissertation* contains many seeds from which Leibniz's philosophy will take its mature form. In particular, in the *Dissertation* Leibniz elaborates a project that can be summarized as follows:

(1) By means of analysis, each concept should be decomposed into its component parts until the first concepts are reached.

(2) Once the first concepts are reached, combine them and produce all kinds of complex concepts.

(3) At the same time choose a system of simple signs to designate the first concepts, so that any complex of signs can be univocally associated with each complex concept.[8]

Leibniz believes that the best signs to employ are *numbers*. If the chosen signs are letters or marks different from numbers, we will have a kind of *universal language*: a language, that is, of pure concepts, accessible to everyone. If numbers (in particular *prime numbers*) are employed to

---

[7] See *DAC*: 1–4.   [8] Cf. *DAC*: 4.

designate the first concepts, then we will have the possibility of transforming each logical argument into a calculus.

After his stay in Paris (1676–9), Leibniz enriches his project for the constitution of the *characteristic art* by the following tasks:

(1) Define a method for developing an analysis of each concept.
(2) Define a method for recombining the first concepts and thus producing all complex concepts.
(3) Define a set of rules for developing a very general calculus based on either the relations of coincidence or of containment holding for terms *and* propositions.

Leibniz assigns the task of realizing points 1 and 2 to a discipline that he calls *general science* (*scientia generalis*) and that he divides into two parts: *analysis* and *synthesis*.[9]

Since the concepts that Leibniz had in mind were those employed in everyday life and in the sciences of his time, if they were to be analysed as required by the constitution of the characteristic art, a kind of general repository was needed to store them (in some order). Leibniz believed that an encyclopedia of all knowledge acquired by mankind during the centuries could play the role of such a 'repository'. Thus, the task of constructing an encyclopedia is integrated into the project for the characteristic art.[10]

Leibniz, however, was uncertain about the structure of the encyclopedia, whether it should be systematic, beginning with principles and axioms and then including all truths that can be derived from the principles, or whether it should contain all items in alphabetical order. Several manuscripts with sketches and projects of the two possible structures clearly show Leibniz's irresolution on this point.[11]

---

[9] On *scientia generalis*, see A VI, 4A: LII–LXXXVII, 352–74, 544; L: 233. On *analysis* and *synthesis*, see L: 173–6, 184–8, 229–34 and Schneider (1970). On the relationships between *scientia generalis* and Leibniz's projects for an encyclopedia of the sciences, see Pelletier (2018).
[10] Cf. *Philosophical Essays*: 8; A VI, 4A: 84, 138, 257, 338–60.
[11] Cf. A VI, 4A: 257, 338–49, 430.

A series of essays written around the same time as the *GI* shows that in this period Leibniz intended to build the universal language on the basis of a very austere grammar that he called *rational grammar* (*grammatica rationalis*).[12] In these essays, Leibniz investigated the grammar of a fragment of Latin (the Latin written and spoken mainly by scientists and philosophers of his time) aiming to reduce it to a limited number of elements. To realize this task, in a long essay entitled *Analysis of Particles* (*Analysis particularum*) Leibniz investigates the behaviour and meaning of several Latin particles.[13] In a text explicitly devoted to *philosophical language* he splits the *terms* (*vocabula*) of the natural language into *words* (*voces*) and *particles* (*particulae*). Words are *nouns, verbs,* and *adverbs*; particles are *prepositions, conjunctions, pronouns,* and even *inflexions* and *cases*. As Leibniz remarks, 'Words constitute the matter, particles the form of the discourse (*oratio*).'[14]

Leibniz's distinction is analogous to that of medieval logicians between *categorematic* and *syncategorematic terms* and roughly corresponds to a more general distinction between *fundamental* (*radicalis*) and *auxiliary* (*servilis*) expressions that Leibniz wants to introduce into *characteristic*.[15] 'Fundamental' or *basic* expressions are *substantives* and *adjectives*; 'auxiliary' expressions are particles. Leibniz characterizes his project as follows:

> Everything in discourse can be analysed into the noun substantive 'being' or 'thing', the copula, i.e. the substantive verb 'is', adjectives, and formal particles.[16]

An important task that Leibniz assigns to *rational grammar* is that of finding a treatment of relational arguments that would permit them to be handled by the methods of what he regards as logic.[17] Statements, and consequently arguments containing relations, indeed, were quite troublesome to people who accepted traditional logic based on Aristotle's syllogistic.

---

[12] Cf. AVI, 4A: 102–5, 112–17, 267, 338–9, 344–5, 528.   [13] A VI, 4A: 646–67.
[14] A VI, 4A: 882.   [15] A VI, 4A: 643.
[16] *LP*: 16 (translation slightly modified); A VI, 4A: 886.   [17] *LP*: xx.

A well-known argument in syllogistic form is the following:

(A)
(1) All men are mortals.
(2) All Greeks are men.
(3) Therefore, all Greeks are mortals.

A typical argument involving relations, instead, is this:

(B)
(1) Socrates is Sophroniscus' son.
(2) Therefore, Sophroniscus is Socrates' father.

According to the mnemonic verses employed by the schoolmen, (A) is a syllogism traditionally classified as an instance of the mode *Barbara*. During the seventeenth century (B) became known as a case of *inversion of relation*.[18] In each sentence of syllogism (A) a predicate is attributed to a subject, whereas the premise and conclusion of the argument based on the inversion of relation state that a relation (son, father) holds between two individuals (Socrates, Sophroniscus). The traditional syllogism employs three terms (in the example above: 'man', 'Greek', and 'mortal') and concludes thanks to the role played by the so-called *middle term* (in the example above: 'man'); the inversion of relation does not have a middle term and was considered, at least by some seventeenth-century logicians, as a *direct (or immediate) inference*. The two arguments are quite different, and it is impossible to express the inversion of relation as a syllogism, maintaining at the same time all the constraints that characterize a syllogism in its traditional form.

Aristotle believed that arithmetic, geometry, optics, and 'in general those sciences which make enquiry about the cause' carry out their demonstrations through the first syllogistic figure.[19] As Jonathan Barnes remarks:

In his *Elements* Euclid first sets down certain primary truths or axioms and then deduces from them a number of secondary truths or theorems. Before ever Euclid wrote, Aristotle had described and commended that

[18] Cf. Jungius (1957): 89–93.   [19] Aristotle, *An. post.* 79a17–24.

rigorous conception of science for which the *Elements* was to provide a perennial paradigm. All sciences, in Aristotle's view, ought to be presented as axiomatic deductive systems—that is a main message of the *Posterior Analytics*. And the deductions which derive the theorems of any science from its axioms must be syllogisms—that is the main message of the *Prior Analytics*.[20]

Aristotle's view was accepted by the great majority of his followers and by those who shared the logical theory developed in *Prior Analytics*. The ancient philosopher and physician Galen of Pergamon (third century AD) was probably the first to claim that the Aristotelian syllogism was unsuitable for handling relations and relational arguments. Galen, indeed, introduced a new class of inferences that he called *relational syllogisms* to handle relations.[21] These syllogisms differed, according to him, from categorical and hypothetical syllogisms:

> There is also another, third, species of syllogism useful for proofs, which I say come about in virtue of something relational, while the Aristotelians are obliged to number them among the predicative syllogisms.[22]

Since Galen, only a very small number of logicians and philosophers have dared to oppose the received view inspired by Aristotle.[23] Among these dissenting voices, that of Joachim Jungius (1587–1657) was one of the most interesting. A professor of natural sciences in Hamburg, the author of a logic handbook, the *Logica Hamburgensis* (1638), and highly regarded by Leibniz, Jungius believed that traditional logic needed to be expanded with additional inferences involving relations that he assumed to be primitive and not reducible to syllogisms.[24]

---

[20] Barnes (2007: 360).   [21] Barnes (2007: 419–24, 431–3).

[22] Galen (1974: xvi, 1).

[23] During the period from around 900 to 1200, some authors belonging to the Arabic tradition were well aware of the difficulties implied by attempting to express relational inferences in the form of an Aristotelian syllogism. From the thirteenth century onwards, Arabic thinkers continued to discuss the problem of relational inferences, still maintaining a logical framework largely inspired by traditional syllogistic doctrines (cf. Khaled El-Rouayheb (2010)), but they seem to have exerted no significant influence on authors belonging to the cultural milieu originating in the Latin tradition.

[24] One of these inferences was the *inversion of relation* just mentioned above; another was the so-called inference *from the right to the oblique* (*a recto ad obliquum*), of which the following is an example:

2.2 Arguments Containing Relations: Rational Grammar and Logic

Relations and relational arguments caused Leibniz two different (but related) kinds of problems: an ontological problem on the one hand, and a logical-syntactic problem on the other. The ontological problem was determined by the basic ontological view prevailing at the time among Western philosophers. According to this view, the entire world was composed of individual beings only, called 'substances', and their inhering accidents. These accidents were thought of as strictly 'monadic', i.e. they could not simultaneously inhere in more than one substance. Yet, since relations link together two or more substances, the problem arises about the nature of 'polyadic' properties corresponding to relations and relational accidents.

The logical-syntactic problem was determined by the structure of the proposition that Leibniz and the great majority of his contemporaries considered basic. For Leibniz the elementary form of any proposition (this too inspired by Aristotle) was *subject-copula-predicate*. As we have seen, however, relational sentences do not have such a form and therefore do not easily conform to the structure of a traditional syllogism.

To solve the first, the ontological problem, Leibniz adopted a strategy analogous to that of Peter Auriol, a medieval thinker who lived about three centuries before him.[25] Leibniz explicitly recognizes the existence of polyadic predicates, but only in so far as they denote 'merely mental

---

Every circle is a figure [*Omnis circulus est figura*];
   Therefore, who describes a circle, describes a figure [*Ergo Quicumque circulum describit, figuram describit*].

This inference was called 'from the right to the oblique' since, whereas in the first premise all terms (*circulus* ['circle'], *figura* ['figure']) are in nominative ('right') case, they are in a case different from nominative ('oblique') in the conclusion (*circulum, figuram* [accusative]). See Jungius (1957: 123).

[25] Peter Auriol (1280–1322) was a French Franciscan who taught at Bologna, Toulouse, and Paris. Concerning the nature of a relation, he wrote in his commentary on the *Sentences*:

a relation is only in apprehension, having no being in things, because that which exists as one and simple and connects [*attingit*] two really distinct things seems to be only the work of the intellect, otherwise the same simple and individual thing will be in several things separated from each other. But it is clear that a relation connects two distinct things, one as a foundation and the other as term. Then, being something indivisible and simple, it cannot be in the extra-mental reality, but only in the consideration of the intellect.

(Auriol, *In I Sent.*, d. 30)

things'.[26] Thus, if Paris loves Helen, the relation of *loving*, according to Leibniz, has a merely mental nature. In the 'real world' there is the couple of individual substances called respectively 'Paris' and 'Helen', each with its internal properties, but there is not a 'real property' connecting them as a 'bridge' and corresponding to 'loving'.

As regards the second problem, which was logical-syntactic in nature, Leibniz proposes a solution strictly connected with the previous one. Given a sentence of the general form 'R(a, b)' stating that a certain relation 'R' holds between two subjects 'a' and 'b', he reduces it to another sentence in which an *auxiliary expression* ('in so far as', 'by that very fact', etc.) connects two sentences in form, such as 'a is P' and 'b is Q', with 'P' and 'Q' each corresponding to one of a pair of correlated terms, such as *lover–beloved*, *murdering–murdered*, and so forth. Thus, in a text entitled *Grammatical Thoughts*, probably written in the autumn of 1678, Leibniz proposes the following analysis of the sentence 'Paris loves Helen':

(a) Paris is a lover, and by that very fact [*et eo ipso*] Helen is a loved one.[27]

Analogously, in the case of 'Caius is killed by Titius':

(b) In so far as Titius is murdering, therefore Caius is murdered.[28]

The two sentences (a) and (b), however, are *not logically equivalent* to the sentences that they are supposed to analyse: (a), for instance, is true in any case in which Paris loves a woman different from Helen and Helen is loved by someone distinct from Paris.[29] And (b) could be true even though Titius does not murder Caius, but someone else, and Caius is murdered by a fourth person.

Leibniz was well aware that there are different types of relations and that the above-mentioned analyses apply mainly to *asymmetrical* relations.[30] He therefore proposes a slightly different treatment for *symmetrical*

[26] Kauppi (1960: 58–60); Mugnai (2012).
[27] A VI, 4A: 114–15. Cf. Mates (1986: 213–18).   [28] A VI 4A: 651.
[29] Leibniz seems to interpret the connectives 'and by this very fact' (*et eo ipso*) and 'in so far as' (*quatenus*) as a kind of 'very strong' conditional, but he does not elaborate this point.
[30] Relations, that is, for which, if it holds that a is R to b, from this it does not follow that b is R to a (i.e. R(a, b) does not imply R(b, a)).

relations. Given a sentence of the general form 'a is similar to b', with 'a' and 'b' names of individuals, he suggests that it should be analysed as 'a is P' and 'b is P', with 'P' designating a property common to both a and b.[31] The rationale behind this analysis (well known to Ockham and other medieval logicians) is that two or more subjects are similar if they share at least a property.[32] In this case too, the truth of 'a is P' and 'b is P' is a necessary condition for the truth of 'a is similar to b', but the latter assertion is not logically equivalent to the conjunction of the first two sentences.

What Leibniz attempts to convey with these grammatical transformations is the idea that relations 'supervene' on the internal properties of two or more subjects belonging to the real world—a result of which we may be aware only by 'thinking together' the related items. Leibniz coins the word 'concogitabilitas' ['co-thinkability'] to express the act of thinking with which we usually grasp relations.[33]

In the *GI*, relations are strongly connected with so-called 'oblique terms' (*termini obliqui*), i.e. terms in a case other than the *nominative*. Thus, Leibniz associates the treatment of oblique terms with that of *partial terms*, such as 'same' or 'similar'. He calls 'integral' a term that may play the role of subject or predicate in a proposition. *Homo* ['man'] and *Caesar*, for instance, are *integral*, whereas *idem* ['same'] and *similis* ['similar'] are *partial*.[34] Leibniz considers as *partial* both proper and common nouns, when they are in a case other than the nominative. The core of Leibniz's distinction may be represented as follows:

Terms

Integral: *Ens* ['Being']; *doctus* ['wise']; *Caesar*; *similis Alexandro* ['similar to Alexander']; *ensis Evandri* ['Evander's sword']...etc.

Partial: *similis* ['similar']; *idem* ['identical']; *Evandri* ['of Evander']; *Alexandri* ['of Alexander']...etc.

---

[31] Cf. A VI, 4A: 11.   [32] Cf. William of Ockham (1974: 281).

[33] Cf. Kauppi (1960: 49); Mugnai (2012).

[34] Obviously, in the proposition '*similar* expresses a relation', the term *similar* plays the role of a subject, even though it is a *partial term*. Here, however, this term is not considered according to its ordinary meaning: it is *mentioned* but not *used*. A medieval logician would say that in this proposition it has 'material supposition'.

Employing a Fregean expression, we may say that Leibniz distinguishes two kinds of terms: the *saturated* and the *unsaturated*. The latter are what he calls 'partial terms', those that are saturated are 'integral terms'. Partial terms cannot enter the logical calculus unless they have been 'saturated'. To saturate a partial term, Leibniz uses an ad hoc linguistic device. The proposition 'Caesar is similar to Alexander' becomes 'Caesar is similar to a thing, which is Alexander'. This is the same logico-linguistic analysis of oblique terms proposed in a letter to Johannes Vagetius (1633–91) in which Leibniz suggests transforming the expression '*qui discit graphicen*' ['he who learns painting'] into the equivalent proposition '*qui discit rem, quae est graphice*' ['he who learns a thing which is painting'].[35] Thus, in the *GI*, Leibniz transforms 'the sword of Evander' into 'the sword, which is a thing of Evander'. In his essays on rational grammar Leibniz moves even a step further, reducing 'the sword, which is a thing of Evander' to 'the sword, which is a thing Evandrian' [*ensis, qui est res Evandria*], in perfect agreement with the fundamental tenet of his logical-grammatical research: not to differentiate between nouns and adjectives.

Since *rational grammar* should make explicit the intimate structure of relational sentences, arranging them in such a way that they can be processed in the logical calculus, it is quite natural that Leibniz should think of *rational grammar* as a prerequisite for logic. As Parkinson remarks:

Leibniz says that logic needs supplementation rather than expansion, and that this supplementation must come from 'rational grammar', which will transform relational arguments into forms which traditional logic can handle.[36]

This explains, the long series of remarks on language and grammar at the beginning of the *GI*, introducing the logical calculus properly speaking.[37]

[35] Vagetius was a pupil of Jungius, and when the latter passed away, he inherited a remarkable quantity of unpublished texts on logic written by Jungius. Vagetius supervised a second edition of Jungius' *Logica Hamburgensis* (1681) and added a long 'Afterword' to it (Vagetius 1977). Leibniz not only read the 'Afterword' and made some remarks on it (A VI, 4B: 1117–21), but even had the opportunity of having a look at Jungius' unpublished papers, the majority of which were later lost in a fire.
[36] *LP*: xx.
[37] Schupp (1982: 144): 'From the point of view of a systematic construction of the sciences, logic comes after rational grammar. This ordering is obeyed even in the *GI*, whose first part (without numbered paragraphs) discusses some basic grammatical distinctions.'

2.3 The Structure of the Proposition:
Extension and Intension

As I observed above, Leibniz assumes that the elementary form of any statement is that of 'subject-copula-predicate', the subject being either a proper noun, like 'Caesar', 'John', or 'Eva', or a common noun, like 'man', 'dog', or 'table'. The predicate is a name of a property, like 'being red', 'being a man', or 'being a philosopher'. *Subject* and *predicate* are *terms*, and a *term* is the conceptual content associated with a Latin noun.[38] The copula expresses the relation of *containment* or *inherence*, which may be considered from two different points of view, in so far as it concerns concepts or individuals (real or merely possible).[39]

In the *Elements of a Calculus*, composed *before* the *GI*, in 1679, the relation of *containment* is introduced as follows:

> every true universal affirmative categorical proposition simply signifies some connection between predicate and subject (a connection based on the nominative case, which is what is always meant here). This connection is that the predicate is said to be in the subject, or to be contained in the subject.[40]

And again in the *Elements* Leibniz explains the difference between the two points of view according to which we can interpret a proposition:

> (11) Two terms which contain each other but do not coincide are commonly called 'genus' and 'species'. These, in so far as they compose concepts or terms (which is how I regard them here) differ as part and whole, in such a way that the concept of the genus is a part and that of the species is a whole, since it is composed of genus and differentia. For example, the concept of gold and the concept of metal differ as part and whole; for in the concept of gold there is contained the concept of metal and something else, e.g. the concept of the heaviest

---

[38] Cf. A VI, 4A: 288 (*LP*: 39): 'By "term" I understand, not a name, but a concept, i.e. that which is signified by a name; you could also call it a notion, an idea.'

[39] Cf. A VI, 6: 486.   [40] *LP*: 17–18 (translation slightly modified); A VI, 4A: 197.

among metals. Consequently, the concept of gold is greater than the concept of metal.

(12) The Scholastics speak differently; for they consider, not concepts, but instances which are brought under universal concepts. So they say that metal is wider than gold, since it contains more species than gold, and if we wish to enumerate the individuals made of gold on the one hand and those made of metal on the other, the latter will be more than the former, which will therefore be contained in the latter as a part in the whole.[41]

Thus, given, for example, the proposition 'Every man is mortal', these are the two different ways of interpreting it:

1. Every individual 'falling under' the concept of *man* belongs to the collection (aggregate, set, or class) of individuals falling under the concept *being mortal*.
2. The concept associated with the word 'man' has among its component parts the concept associated with the word 'animal'.

In the *New Essays,* more than twenty years after the *Elements,* Leibniz comes back to the difference between the two points of view and gives a name to each of them:

This manner of statement deserves respect; for indeed the predicate is in the subject, or rather the idea of the predicate is included in the idea of the subject.... The common manner of statement concerns individuals, whereas Aristotle's refers rather to ideas or universals. For when I say *Every man is an animal,* I mean that all the men are included amongst all the animals; but at the same time I mean that the idea of animal is included in the idea of man. 'Animal' comprises more individuals than 'man' does, but 'man' comprises more ideas or more attributes: one has more instances, the other more degrees of reality; one has the greater extension, the other the greater intension.[42]

---

[41] *LP:* 20 (A VI, 4A: 199–200).
[42] *NE:* 486. On the distinction between intension and extension, see Kauppi (1960: 7–12); Lenzen (1983); Swoyer (1995).

Of the two approaches, that based on *intension* and that focused on *extension*, Leibniz definitely opts for the first, and in the *Elements of a Calculus* explains the reason for his choice: 'However, I have preferred to consider universal concepts, i.e. ideas, and their combinations, as they do not depend on the existence of individuals.'[43]

Leibniz firmly believed that the relationships between *intension* and *extension* were ruled by a *Principle of Reciprocity* according to which, if a given concept B is contained in the intension of the concept A, then the extension corresponding to A is contained in the extension corresponding to the concept B.[44] As Leibniz remarks in an essay dated 'August 1690':

> the method which proceeds by means of notions is the contrary of that proceeding by means of individuals. If all men, indeed, are part of all animals, that is, if all men are in all animals, vice versa the notion of animal will be in the notion of man; and if there are more animals besides men, something needs to be joined to the idea of animal to obtain the idea of man: increasing the number of conditions, the number [of individuals] decreases.[45]

If we interpret the *extension* associated with each notion as the *set* of individuals subordinated to it, it is quite natural to represent the relation of containment (or inherence) considered from the *extensional point of view* as equivalent to the *set-theoretical inclusion*. Thus, given any two terms 'A' and 'B' whatsoever, we may express the fact that the extension of A is included in that of B as follows:

$$_{Ext}A \subset {}_{Ext}B.^{46}$$

On the other hand, if we employ the symbol '>' to represent the relation of containment between notions or ideas, we may represent the fact that the term A contains the term B as follows:

$$_{Int}A > {}_{Int}B.$$

---

[43] *LP*: 20 (A VI, 4A: 200).
[44] Cf. Lenzen (1983: 140–2); Lenzen (2004b: 10–12).   [45] *Opuscules*: 235.
[46] Lenzen (1983: 141).

Therefore, the *Principle of Reciprocity* (PR) between *extension* and *intension* of two terms A and B assumes the form:

$$_{Ext}A \subset _{Ext}B \text{ if and only if } _{Int}A > _{Int}B.$$

So expressed, (PR) is composed of two halves:

(PR.1) If $_{Ext}A \subset _{Ext}B$, then $_{Int}A > _{Int}B$.
(PR.2) If $_{Int}A > _{Int}B$, then $_{Ext}A \subset _{Ext}B$.

(PR.2) does not create problems and seems to be quite reasonable: if concept B is included in concept A, then the inverse relation holds between the extensions of A and B. Things, however, are different with (PR.1): if the extension of a concept A is included in the extension of a concept B, from this it does not follow that the intension of B is included in that of A. A classic counterexample is Quine's famous case of *cordate* and *renate*. In our world, all animals with a *heart* (*cordate*) are animals with *kidneys* (*renate*), but the two concepts of *having a heart* and *having kidneys* are quite distinct.[47] A standard solution of this problem is to assume that the extensions of our concepts are independent of the actual status of our world. Thus, suppose that U is the set of all possible individuals and ϕ a function with U as domain and the set of all concepts as co-domain, and that ϕ assigns a set of elements of U to each concept in such a way that if two concepts A and B differ in intension, then their extensions differ as well. Quine's counterexample is avoided, because, even though in our world all individuals with a heart have kidneys, there are *possible individuals* in U which have a heart but not kidneys (or have kidneys but not a heart). Therefore, in this case (PR.1) holds too. This solution is analogous to that proposed by Leibniz when he writes in the *Elements of a Calculus* that he prefers 'to consider universal concepts, i.e. ideas, and their combinations', instead of extensions, because they 'do not depend on the existence of individuals' (in other words, he considers not only existing but also *possible* individuals).[48]

---

[47] Cf. Lenzen (1983: 141); Swoyer (1995: 103); Lenzen (2004: 11).
[48] Cf. n. 18 above and *LP*: 20 (A VI, 4A: 200).

2.4 Coincidence, Containment, and Substitutivity

Independently of the approach chosen, in favour of either *intension* or *extension*, in the *GI* Leibniz recognizes two properties of the relation of containment, *reflexivity* and *antisymmetry*:[49]

(37) B is B, for B = B (by 10). Therefore B is B (by 36).[50]
(30) That A is B and B is A is the same as that A and B coincide.[51]

Since Leibniz uses 'A is B' as synonymous with 'A contains B', the transitivity of containment is stated as follows in the *GI*:

(19)... If, for instance, A is B and B is C, A will be C.[52]

Thus, the relation of containment imposes a *partial order* on the intended domain of terms of the artificial language (presupposed by the *GI*).

Leibniz, however, in the *GI*, assumes the relation of *coincidence* as *basic* and expresses containment by means of this relation:[53]

(83) In general, that A is B is the same as A = AB: from this, indeed, it is clear that B is contained in A and that 'man' and 'man animal' are the same.[54]

---

[49] In the 1679 essays on logical calculus, Leibniz takes containment as a primitive relation, assuming reflexivity and transitivity as, respectively, a 'proposition true in itself' and an 'inference true in itself' (cf. *LP*: 33).

[50] *GI*: 63 (A VI, 4A: 754).

[51] *GI*: 63 (A VI, 4: 753). Cf. also A VI, 4A: 839: 'If A is in B and B is in A, then A = B'; Lenzen (2004a: 283–4).

[52] *GI*: 7 (A VI, 4A: 752). As Malink and Vasudevan (2016: 706) remark, Leibniz gives a proof of the transitivity of inclusion in *Opuscules*: 229–30.

[53] At (30) Leibniz characterizes coincidence as mutual containment: 'That A is B and B is A is the same as that A and B coincide.' On the relationship between *containment* and *coincidence* in Leibniz's logical essays, see Malink and Vasudevan (2019).

[54] *GI*: 69 (A VI, 4A: 765). Cf. Malink and Vasudevan (2016: 692):

Leibniz is here appealing to the intuition that if A coincides with some kind of B, then one specific kind of B with which A coincides is the composite term AB. Note that this definition of containment relies on the fact that the terms of Leibniz's calculus are closed under the operation of composition. Thus, by positing this definition, Leibniz manages to decrease the number of primitive relations that obtain between terms by increasing the syntactic complexity of the terms themselves. In thereby shifting the focus from the relational structure of the terms to their compositional

Containment was closely related by Leibniz to the truth of a proposition and, as we will see in Section 3, this connection caused serious problems for his metaphysics. For the moment, however, let me simply recall Leibniz's claim about the characteristic feature of all true (affirmative) propositions:

> An affirmation is true if its predicate is in its subject; thus, in every true affirmative proposition, necessary or contingent, universal or singular, the concept of the predicate is somehow contained in the concept of the subject, in such a way that anyone who understood the two concepts as God understands them would *eo ipso* perceive that the predicate is in the subject.[55]

Here, Leibniz explicitly states that the analytical criterion for truth holds for *any kind* of affirmative proposition, *necessary* or *contingent*, universal or singular, thus disconnecting *necessity* from *analyticity*.

As regards *coincidence*, Leibniz recognizes that it enjoys the properties of *reflexivity*, *symmetry*, and *transitivity*:

> (10) A *proposition true in itself* is 'A coincides with A'[56]
> (6) If A coincides with B, B coincides with A.[57]
> (8) If A coincides with B and B coincides with C, then A also coincides with C.[58]

Therefore, the relation of coincidence is an *equivalence relation*.

Leibniz usually employs the principle of substitutivity to define *coincidence*. Sometimes, however, he has recourse to mutual containment to attain the same result. Thus, in the *Elements of a Calculus* (1679), we have:

> (10) Two terms which contain each other and are nevertheless equal I call 'coincident'. For example, the concept of a triangle coincides in effect with the concept of a trilateral—i.e. as much is contained in the one as in

---

or algebraic structure, Leibniz's reduction of containment to coincidence marks a significant advance in the algebraic treatment of logic.
[55] MP: 96.   [56] GI: 61 (A VI, 4A: 751).   [57] GI: 61 (A VI, 4A: 750).
[58] Ibid.

the other. Sometimes this may not appear at first sight, but if one analyses each of the two, one will at last come to the same.[59]

And in the *GI*:

(30) That A is B and B is A is the same as that A and B coincide, i.e. that A coincides with B, which coincides with A.[60]

The substitutivity principle is one of the basic principles of Leibniz's logical calculi, and this is how Leibniz presents it in an essay of 1679:

(7) Those terms are the same of which one can be substituted in place of the other without loss of truth, such as 'triangle' and 'trilateral', 'quadrangle' and 'quadrilateral'.[61]

Leibniz states the principle of substitutivity at the beginning and at the end of the *GI*, when he resumes the main principles of his calculus:

That A is the same as B means that one can be substituted for the other in any proposition whatsoever without loss of truth.[62]

and

(198) Principles:
1st. ⟨Coincidents can be substituted for one another.⟩[63]

The following passage from the *GI* shows that Leibniz was well aware of the existence of contexts in which the substitutivity principle fails, contexts, that is, that we now call 'intensional', 'opaque', or 'oblique':

---

[59] *LP*: 20. Parkinson's translation is faithful to Leibniz's text but the English word 'nevertheless' (which correctly translates the Latin *nihilominus*) sounds odd, because two terms that contain each other *necessarily* coincide. As Lenzen suggests (Lenzen (2019: 199, n. 5)) Leibniz probably had in mind the fact that if a given term B is contained in a given term A, then A and B differ as a *part* from the *whole*.

[60] *GI*: 63 (A VI, 4A: 753).   [61] Ibid. Translation slightly modified.
[62] *GI*: 59 (A VI, 4A: 746).   [63] *GI*: 81 (A VI, 4A: 786).

things that coincide, indeed, can be substituted for one another
(except in the case of propositions which you could call 'formal', where
one of the coinciding things is taken formally in such a way that it
is distinguished from the others; these propositions, however, are
reflexive, and are asserted not so much about a thing as about our way
of conceiving it, where there is certainly a distinction between them).[64]

The same idea is expressed in an essay on geometry:

I define as the same [eadem] those things that can be substituted everywhere for one another *salva veritate*, in those propositions, I mean, that
are direct and do not refer to a way of considering the subject itself.[65]

In an essay probably written some years before the *GI*, entitled *General
Remarks (Notationes generales)* Leibniz gives an example of such *reflexive* or *indirect* contexts. Even though *Peter* and *the Apostle who denied
Christ* denote the same person, if we substitute 'Peter' for 'the Apostle
who denied Christ' in the sentence 'Peter, in so far as he was the Apostle
who denied Christ, committed a sin', we obtain a new sentence with a
different truth value. Peter, indeed, did not commit a sin in so far as he
was Peter. As Leibniz remarks, the subject of this proposition is considered in a particular respect (that of having denied Christ) and not
simply as Peter.[66]

2.5 Privation and Composition

Besides the principle of substitution and the relation of coincidence
(and containment), other ingredients of Leibniz's logical calculus in
the *GI* are:

the unary operation of 'privation';
the binary operation of *composition*.

[64] *GI*: 62 (A VI, 4A: 752). A similar passage is found at A VI, 4A: 672.
[65] In Mugnai (1992: 147).   [66] A VI, 4A: 552.

Both these operations apply to *terms*, which are designated by means of letters of the Latin alphabet. As usual in the mathematical practice of the time, Leibniz employs the first letters to designate terms with a *definite*, known meaning and the last to designate terms with an *indefinite*, unknown meaning. Consequently, he speaks of *definite* and *indefinite letters* (which play a role analogous to that of *variables* in contemporary logic).

Leibniz states some rules concerning the indefinite letters that are worth noting, in particular those at §§20, 23, and 24:

> (20) Something should be noted here that should have been stated earlier in this calculus: that one letter can be put in place of any number of other letters together; for example: YZ = X. But this has not yet been employed in this calculus of reason, lest confusion arise.
>
> (23) An indefinite letter not yet used can be substituted for any definite letter. Similarly, one can perform the substitution for any number of definite letters, and for definite and indefinite ones as well, i.e. one can put A = Y.
>
> (24) To any letter a new indefinite one can be added; thus, for A we can put AY, because AA = A (by 18) and A is Y (or for A we may put Y, by 23). Therefore A = AY.[67]

Indefinite letters were employed by Leibniz in essays written *before* the *GI*, to express the categorical propositions of traditional syllogistic in the form of algebraic equations. Given a proposition such as 'every man is an animal', he proposes first to reduce it to '(Every) *m* is *a*' (where '*m*' = man and '*a*' = animal) and then represent the proposition so transformed as an equation: '*m* = *xa*'.[68] The indefinite letter '*x*' here plays the role of the *difference* 'rational' that, juxtaposed with 'animal', gives rise to the composite term 'rational animal'. In the *GI* we find the same idea associated with the further remark that an indefinite letter in cases like this plays the role of the quantifier 'some':

---

[67] *GI*: 62 (A VI, 4A: 752).
[68] Cf., for instance, A VI, 4A: 209: 'Thus, any proposition can always be transformed into an equation.'

(16) An *Affirmative proposition* is 'A is B' or 'A contains B' or, as Aristotle says, 'B is in A' (i.e. in the nominative case).That is, if we substitute a value for A, we will have: 'A coincides with BY'. For example: 'man is an animal', i.e. 'man is the same as animal...', namely 'man is the same as a rational animal'. With the sign 'Y', indeed, I mean something undetermined, so that 'BY' is the same as 'some B' or '...animal' (where 'rational' is implied, provided that we know what is to be implied) or 'some animal'. Thus, 'A is B' is the same as 'A is coincident with some B' or A = BY.[69]

Leibniz thinks of employing the indefinite letters even for representing the universal quantifier 'Every', but he does not develop any systematic account of quantification by means of indefinite letters.[70]

Privation is a unary operation that Leibniz designates by means of the Latin word *non*. Applied to the term, 'man' for instance, it produces the privative term 'non-man', i.e. the *privative* of 'man'. *Privation* behaves like classical *negation*:[71]

(96) Non-non A = A.[72]
(2) If A and B coincide, then non-A and non-B also coincide.... Non-non-A and A coincide; thus, if non-A and B coincide, non-B and A will also coincide.[73]
(4)... 'Non' applied to 'non' is equivalent to the omission of both.[74]
(93) If A is B, non-B is non-A.[75]
(94) If non-B is non-A, A is B.[76]

To express *composition*, Leibniz does not employ any particle or specific logical symbol. He simply juxtaposes the letters representing terms (concepts) giving rise to a composition. Thus, for example, the term 'AB' is the composite term obtained by juxtaposing the terms corresponding

---

[69] *GI*: 62 (A VI, 4A: 751).
[70] On Leibniz's various attempts to develop a theory of quantification, see Lenzen (1984a) and (2004b: 47–73).
[71] On Leibniz's theory of negation, see Lenzen (1986).   [72] *GI*: 67 (A VI, 4A: 767).
[73] *GI*: 60–1 (A VI, 4A: 749).   [74] *GI*: 61 (A VI, 4A: 750).
[75] *GI*: 70 (A VI, 4A: 766).   [76] *GI*: 70 (A VI, 4A: 767).

to the letters 'A' and 'B'; 'ABC' is the composite term obtained by juxtaposing the letters 'AB', and 'C', and so on. For *composition* Leibniz explicitly recognizes the properties of *commutativity*:

(147) ... AB being the same as BA ...[77]

and *idempotence*:

(16) ... This notation, however, presupposes that AA is the same as A, for a redundancy arises.[78]

(18) Due to the nature of this characteristic A, AA and AAA, etc., i.e. 'man' and 'man man' and 'man man man' coincide.[79]

As for *associativity*, Leibniz employs it only in few cases in the *GI*, but does not state it explicitly.[80]

2.6 Logic of Terms and Logic of Propositions: Some Algebraic Interpretations

One of the most interesting achievements of the *GI* is the attempt to develop a logical calculus for both terms *and* propositions based on the logic of terms. In other words, Leibniz attempts to develop the logic of propositions on the basis of the logic of terms. These are the relevant passages of the *GI* where Leibniz alludes to his project of reducing hypothetical propositions to categorical ones:

(75) If, as I hope, I can conceive all propositions as terms, and all hypotheticals as categoricals, and if I can treat them all in the same way, this promises a wonderful facility in my characteristic art and analysis of concepts, and it will be a discovery of the greatest importance. Clearly, in

---

[77] *GI*: 78 (A VI, 4A: 780).   [78] *GI*: 62 (A VI, 4A: 751).
[79] Ibid. This corresponds to George Boole's *index law*; cf. Boole (1847: 17–18).
[80] Cf. Castañeda (1976: 488); Lenzen (1984a: 201); Peckhaus (1997: 48); Hailperin (2004: 327). *Associativity* was widely employed by mathematicians in Leibniz's time, but it was only in the first half of the nineteenth century that it began to be explicitly mentioned as a relevant property of some operations. On the relationships between commutativity of composition and the principle of associativity in the *GI*, see Malink and Vasudevan (2016: 699).

general, I call a term 'false' which in the case of incomplex terms is an impossible term, or at least a meaningless one, and which in complexes is an impossible proposition, or at least a proposition which cannot be proved. Thus the analogy remains. So, by 'A' I understand either an incomplex term or a proposition or a collection or a collection of collections, etc.; so that, in general, a term is true which can be perfectly understood.[81]

(137) Therefore we have disclosed many secrets of great importance for the analysis of all our thoughts, and for the discovery and demonstration of truths. We have discovered how all truths can be explained by numbers; how contingent truths arise and that they have, in a certain sense, the nature of incommensurable numbers; how absolute and hypothetical truths have one and the same laws and are contained in the same general theorems, so that all syllogisms become categorical. Finally, we have discovered what the origin of abstract terms is, and now it will be worth our while to explain this last point a little more clearly.[82]

(189) The principles, therefore, will be these:

...

Sixth, Whatever is said of a term containing a term can also be said of a proposition from which another proposition follows.[83]

To realize his project, Leibniz has recourse to what he calls 'logical abstract terms' (*abstracta logica*), which, in a text entitled *On the abstract and the concrete*, written after the *General Inquiries*, he proudly claims to have introduced in philosophy.[84]

In *On the abstract and the concrete*, Leibniz distinguishes two kinds of abstract terms: *metaphysical* and *logical*. *Metaphysical abstract terms* are those of traditional scholastic philosophy, like *redness, humanity,* or *rationality*, which were supposed to denote universal properties constituting the concrete things of the world. *Logical abstract terms*, on the other hand, 'are subsequent' to concrete things.[85] In the same text,

---

[81] *GI*: 68 (A VI, 4A: 764).   [82] *GI*: 76 (A VI, 4A: 777).
[83] *GI*: 76 (A VI, 4A: 784).   [84] Cf. A VI, 4A: 987–94.
[85] A VI, 4A: 992. *Wisdom*, being a substantive, may suggest that it refers to something existing 'by itself', independently of the subjects that are wise; *being wise*, instead, refers to a subject of which it is a modification and suggests that it cannot exist independently of this subject.

Leibniz states that the abstract terms of the scholastic tradition and his logical abstract terms maintain the same relationship with concrete things: as some learned man has wisdom, so he has the property of *being wise*. *Being wise* (*esse sapientem*), *being just* (*esse iustum*) are instances of *logical abstract terms*. It is thanks to the abstraction performed by 'logical abstract terms', that any proposition of the form '*A* is *B*' ('*A* contains *B*') gives rise to the new term *A's being B* or *the B-ness of A*.[86]

In the text on the abstract and the concrete, Leibniz writes that the invention of the *logical abstract terms* is in accordance with Horace's saying that *virtue amounts to avoiding vice*—the vice being, in this case, any kind of strong ontological commitment.[87] Thus, it is no wonder that in another short essay from the same period Leibniz defines himself as a 'nominalist at least for precautionary reasons [*eatenus sum nominalis, saltem per provisionem*]'.[88] Consequently, he considers 'the abstract terms not as corresponding to things, but as a kind of shorthand for the discourse'.[89]

Having introduced *logical abstract terms*, Leibniz employs them to transform conditionals into categorical propositions. Given a conditional of the form 'If *A* is *B*, then *C* is *D*', Leibniz suggests transforming it into: '*A's being B is (or contains) C's being D*'. This is the passage of the *General Inquiries* where Leibniz proposes the transformation:

> (138) For if the proposition '<u>A is B</u>' is considered as a term, as we have explained that it can be, there arises an abstract term, namely 'A's being B'. And if from the proposition 'A is B' the proposition 'C is D' follows, then <u>from this there comes about a new proposition</u> of the following kind: '<u>A's being B</u> is or contains <u>C's being D</u>', i.e. 'the B-ness of A contains the D-ness of C' or 'the B-ness of A is the D-ness of C'.[90]

Thus, using the device of *logical abstract terms* and the relation of containment, Leibniz reduces 'consequences to propositions, and propositions to terms':

---

[86] Ibid.   [87] Ibid.; cf. Hor., *Epist.*, I, 1, 41.
[88] A VI, 4A: 996.   [89] Ibid.
[90] *GI*: 76 (A VI, 4A: 777). The superscript lines in the quotation serve as a bracketing device.

8th. That a proposition follows from a proposition is nothing else than that the consequent is contained in the antecedent, as a term in a term, and by this method we reduce consequences to propositions, and propositions to terms.[91]

The same claim is repeated in the essay on the abstract and the concrete:

as I reduce [*reduco*] categorical propositions to simple terms modified by 'is', so I reduce hypothetical to categorical propositions in which those abstract terms occur. So, I reduce this hypothetical proposition: *If Peter is wise, then Peter is just* to this categorical: *Peter being wise is Peter being just*; thus, the same rules hold for hypothetical as for categorical propositions.[92]

Leibniz distinguishes *coincidence*, when it holds between terms, from coincidence holding between sentences:

Any letter, such as 'A', 'B', or 'L' means for me either some integral term or another integral proposition.

When a single term is put for several ones, the latter are the definition or the assumed value, and the former is the defined value, for example: if for AB I put C, i.e. when A = BC is a primitive proposition.

A and B coincide if, by substitution of the assumed values in place of the terms and conversely, the same thing, i.e. the same formula, appears on both sides.

I say that statements *coincide* if one can be substituted for the other without loss of truth, i.e. if they infer each other reciprocally.[93]

That the invention of the *logical abstract terms* is neither accidental nor restricted to the logical writings is clearly shown by what Leibniz says in the *New Essays*, almost twenty years after the *General Inquiries* (and the text on the abstract and the concrete):

---

[91] *GI*: 27 (A VI, 4A: 787).
[92] A VI, 4A: 992.   [93] *GI*: 60 (A VI, 4A: 748).

Now, I always distinguish two sorts of abstract terms: *logical* and *real*. *Real abstract* terms, or at least those which are conceived as real, are either essences or parts of an essence, or else accidents—i.e. beings added to a substance. *Logical abstract* terms are predications reduced to single terms—as I might say 'to-be-man', 'to-be-animal'—and taken in this way we can assert one of the other: 'To be man is to be animal'. But with realities we cannot do this. We cannot say that *humanity* or *manness* if you like, which is the whole essence of man, is *animality*, which is only a part of that essence.[94]

Thus, Leibniz's *logical abstract terms* perform two distinct but coordinate tasks: on the one hand, they undermine the ontological nature of the abstract entities of the scholastic (realistic) tradition and, on the other hand, they generate those terms that Marko Malink and Anubav Vasudevan appropriately label *propositional terms*, making it possible to reduce the calculus of propositions to the calculus of terms.[95] This means that the logical abstract terms can be considered as a kind of 'bridge' between logic and ontology (or metaphysics).

As I remarked above, the relation of containment, being reflexive, anti-symmetrical, and transitive, imposes a *partial order* on the set of terms of Leibniz's calculus. Moreover, given two terms A and B whatsoever, juxtaposing them we have:

$$(1^*) \ AB \subset A,$$
$$(2^*) \ AB \subset B,$$

i.e. 'the composite term "AB" is a *lower bound* of both A and B'.[96] Thanks to the principle of substitution, we may determine that AB 'is an upper bound of any lower bound of both A and B'; therefore, it is the *greatest lower bound* of A and B.[97] In other words, we may interpret algebraically the set of terms of Leibniz's logical calculus as a *semi-lattice with meet*.

[94] *NE*: 333–4.
[95] Malink and Vasudevan (2019: 20, n. 49), claim that they have borrowed the label 'propositional term' from Sommers (1982: 156), (1993: 172–3). As they observe, the same expression occurs in Barnes (1983: 315) and Swoyer (1995: 110–11).
[96] Malink-Vasudevan (2016: 707).   [97] Ibid.

If we add the operation of *negation* (corresponding to Leibniz's *privation*) to this semi-lattice with meet, we get a *lattice*. In a *meet semi-lattice*, indeed, it is possible to introduce the operation of *join* (corresponding to the Boolean *union*) by means of one of De Morgan's laws.[98] Thus, we have all the ingredients of a Boolean algebra in the proper sense.

To realize the project of reducing the logic of propositions to the logic of terms, Leibniz employs three principles:

(1) the principle of *bivalence*, according to which each proposition is either true or false;

(2) the principle of propositional negation;

(3) the principle of propositional containment.

The principle of bivalence (together with the principle of non-contradiction) is asserted very early in the *GI*:

(4)...From these it is also demonstrated that every proposition is either true or false, i.e. if L is not true, it is false; if it is true, it is not false; if it is not false, it is true; if it is false, it is not true.[99]

The principle of propositional negation explains that if we put negation (*privation*) in front of a proposition, this means that the proposition (better: the corresponding *propositional term*) is *false*:

(32)...If B is a proposition, then 'non-B is the same as B' is false, i.e. B's being false.[100]

Finally, propositional containment reduces the relation of 'following from...' subsisting between propositions to the relation of containment between propositional terms.

---

[98] In the case of sets, the so-called De Morgan's laws relate the set-theoretical operations of intersection and union by means of the operation of complementation. Thus, for instance, the intersection of a set A with a set B is logically equivalent to the complement of the union of, respectively, the complement of A and the complement of B. In the propositional case, De Morgan's laws relate conjunction and disjunction through negation. Thus, the conjunction of two sentences p and q is logically equivalent to the negation of the disjunction of the negations of, respectively, p and q.

[99] *GI*: 61 (A VI, 4A: 750).   [100] *GI*: 63 (A VI, 4A: 753).

As Marko Malink and Anubav Vasudevan have shown, it is possible to interpret Leibniz's logical calculus as it is developed in the *GI* as a particular kind of algebraic structure that they call 'auto-Boolean algebra'. A peculiar feature of such an algebra is that it takes seriously the role played by propositional terms in the Leibnizian calculus and offers a very simple (and elegant) solution to the problem of reducing the logic of propositions to the logic of terms.[101]

At this point, however, some explanations are in order. That the logical calculus of the *GI* contains all the ingredients for developing a Boolean algebra was shown by Wolfgang Lenzen (1984b). Leibniz, however, seems to ignore the property of *duality* that in his calculus subsists between the operations of *conjunction* and *disjunction*. Obviously, he knows the so-called De Morgan's laws that link conjunction (*et*—'and') and disjunction (*vel*—'or') by means of *negation* (they were recognized by almost all the scholastic logicians), but he does not give special emphasis to them and, moreover, as far as we can judge on the basis of the texts published up to this point, the logical operation of disjunction is conspicuously absent from his calculi. This sounds particularly odd, if we consider that Leibniz was probably the first to propose the symbol '∨' to represent the logical (non-exclusive) disjunction and that Giuseppe Peano took this symbol from a text published by Gerhardt:

> As '+' is a conjunctive sign, i.e. a sign of composition corresponding to 'and' (so that 'a + b' is a and b simultaneously), so there is a disjunctive sign, that is, a sign which means an alternative, corresponding to [the Latin word] 'vel'. Thus, to me 'a ∨ b' means 'a or b'.[102]

---

[101] Cf. Malink-Vasudevan (2016: 712–13):

Leibniz's calculus can…be viewed as a calculus for reasoning about terms insofar as they constitute an auto-Boolean algebra. The distinctive feature of such an algebra is a binary operation which determines the place in the algebra of the propositional term 'A = B' as a function of the terms A and B. This operation, which maps any two coincident terms to 1 and any two noncoincident terms to 0, is not Boolean. That is to say, except in the degenerate case in which every term coincides with either 0 or 1, this operation cannot be defined by means of composition and privation alone. Accordingly, the methods of reasoning licensed by the propositional principles of Leibniz's calculus constitute a genuine extension of Boolean reasoning.

[102] Cf. GM 7: 57.

Leibniz's behaviour towards logical disjunction is even more puzzling if we consider that he attributes to it the property of *idempotence*:

> Not all formulae signify a quantity and we may find an infinity of ways to perform the calculus. In case of a calculus based on disjunction [*pro calculo alternativo*], for instance, if one says that *x* is *abc*, one may understand that *x* is *a* or *b* or *c*.... In this calculus, indeed, *a* and *aa* are equivalent and any combination whatsoever of a letter with itself does not matter at all.[103]

This is a casual remark that, as far as I know, Leibniz did not develop further, as he did not work out a logical calculus based on disjunction.

3 Metaphysics

3.1 The Main Puzzle

According to the scholastic tradition which may be traced back to Duns Scotus, Leibniz defines the logically possible as that which does not imply a contradiction: 'If something implies a contradiction, it is called impossible, otherwise it will be understood as possible.'[104] The following definitions, simply repeat a doctrine that was quite common amongst medieval and late medieval thinkers:

> Impossible is what implies an absurdity;
> Possible is what is not impossible;...
> Necessary is that whose opposite is impossible;
> Contingent is what is not necessary.[105]

As we have seen, Leibniz extends his containment theory of truth to any kind of propositions, including both the contingent and the necessary.

---

[103] A VI, 4A: 511–12. This constitutes a remarkable improvement even on Boole's original system, which did not consider the operation of union (join) as idempotent; cf. Hailperin (1981).

[104] A VI, 4A: 626. Cf. Duns Scotus' definition: 'The *possibile logicum* is the mode of the composition formed by the understanding, whose terms do not include any contradiction' (*Opus Oxoniense*, I, dist. 2, *quaest.* 6, *art.* 2, *art.* 10).

[105] LH IV, 7c, Bl. 78r.

In doing so, however, he separates *analyticity* from *necessity*. He admits, indeed, that there are propositions that, even though analytically true, are not necessary (the contingent ones). Conversely, he establishes a link between analyticity and contingency, because every contingent truth, being a truth, is perforce analytical. Thus, Leibniz has to face the problem of explaining how a proposition can be analytically true and contingent at the same time. If the proposition 'Caesar crossed the Rubicon' is true, it is *analytically true* (the concept of crossing the Rubicon is included in Caesar's concept), but then, how can it be *contingent*?[106]

Leibniz discussed this problem at length in his correspondence with the French philosopher Antoine Arnauld. It was Arnauld, indeed, who was the first to emphasize the odd consequences deriving from putting together, as Leibniz does, the predicate-in-the subject theory of truth with the theory of the complete concept.[107] In the *Discourse on Metaphysics* (written at the same period as the *GI*), Leibniz connects these two theories as follows:

> Now it is evident that all true predication has some basis in the nature of things and that, when a proposition is not an identity, that is, when the predicate is not explicitly contained in the subject, it must be contained in it virtually. That is what the philosophers call *in-esse*, when they say that the predicate is in the subject. Thus, the subject term must always contain the predicate term, so that one who understands perfectly the notion of the subject would also know that the predicate belongs to it. Since this is so, we can say that the nature of an individual substance or of a complete being is to have a notion so complete that it is sufficient to contain, and to allow us to deduce from it, all the predicates of the subject to which this notion is attributed.[108]

Arnauld reacted to this, remarking that if Leibniz were right, then everything would be 'obliged to happen through a more than fatal necessity'.[109] To see why, consider the true proposition 'Caesar crossed

---

[106] According to Leibniz, *contingency* is a necessary condition for a human action to be free. On the relationships between *contingency* and *freedom*, see McDonough (2018).

[107] Cf. Di Bella (2018).   [108] Leibniz, *Philosophical Essays*: 41.

[109] GP 2: 15. On the dispute between Leibniz and Arnauld concerning the necessity of human actions, see Sleigh (1990).

the Rubicon': since this proposition is true, the predicate 'crossing the Rubicon' must be 'virtually' contained in Caesar's complete concept; but then the proposition, being analytically true, should be necessarily true. Even though Leibniz is credited with having coined the definition according to which *a proposition is necessary if it is true in all possible worlds*, there is no trace of it in the texts that have been published up to this point. Or, to put it less strongly, there is no trace of it *in this explicit form*, as a direct, plain definition of a necessary proposition.[110] It was Benson Mates, however, who pointed out that at least one paper exists in which Leibniz seems to come very close to the definition that is now standard in the so-called possible-world semantics. The paper in question, belonging to the years around 1686, is devoted to explaining the nature of some crucial notions, such as those of truth, freedom, and contingency, and contains the following passage:

> Essential are those truths that can be demonstrated by analysis of terms, i.e. those that are necessary or virtually identical, the opposite of which is impossible, i.e. virtually contradictory. And such propositions are eternally true, and will not only obtain while the World remains, but would even have obtained if God had created the World on a different plan.[111]

In his monograph on Leibniz, Robert Adams observes that what Leibniz says here 'is suggestive, but not conclusive' as evidence of the fact that he 'conceived of necessity as truth in all possible worlds and contingency as truth in some but not all possible worlds'.[112] According to Adams, indeed, 'on any reasonable interpretation Leibniz regards no world as possible in which something demonstrable is false', but what 'we want to know is whether he thought that all the truths that do not depend on which world God created are necessary'.[113] On this point, however, as Adams remarks, Leibniz seems to be quite unclear.

Even though Leibniz never asserted explicitly that a sentence is necessary if it is true in all possible worlds, he is surely the first to elaborate a

---

[110] This 'negative' remark was made first by Kauppi (1960: 247) and then was reinforced by Mates (1972: 87) and (1986: 107).
[111] MP: 98 (A VI, 4B: 1517).   [112] Adams (1994: 46).   [113] Ibid.

metaphysics in which a complex theory of possible worlds is systematically connected with the modal concepts of *possibility, necessity,* and *contingency.* Clearly, it would be anachronistic to attribute to Leibniz the discovery of the possible-world semantics, but we cannot deny that his philosophical work, even confusedly, foreshadows some aspects of this semantics. At any rate, I do not intend to insist on this point: my aim is that of shedding some light on Leibniz's main puzzle on contingency, employing, as far as possible, 'conceptual tools' explicitly admitted and recognized by Leibniz himself.

In other words, how can a proposition be *analytically but not necessarily true?* Leibniz offers two different solutions to this problem:

(1) a solution based on the so-called 'divine decrees;
(2) a solution that rests on the notion of *infinite analysis.*

Let me now examine each of them in the given order.

3.2 Solution 1

According to Leibniz, amongst the components of a complete concept there are some general principles or decrees determining some essential features of the corresponding individual substance and therefore of the world to which it belongs. These decrees depend on God's will:

> But the concepts of individual substances, which are complete and suffice to distinguish their subjects completely, and which consequently involve contingent truths or truths of fact, and individual circumstances of time, place, etc., must also involve in their concept taken as possible, the free decrees of God, also viewed as possible, because these free decrees are the principal sources of existences or facts. Essences, on the other hand, are in the divine understanding prior to any consideration of the will.[114]

---

[114] A II, 2: 71; L: 332.

A typical divine decree, for example, is the principle according to which the behaviour of all human beings belonging to the actual world must be determined by choosing what they judge to be their good.[115] Now, to embody this decree (as merely possible), in the complete concept of each human being, depends on God's will only. Had God embodied in human beings, and therefore in the complete concept of Caesar as well, a different principle of action, Caesar could not have crossed the Rubicon. In this sense, the property of *having crossed the Rubicon*, so argues Leibniz, is a contingent property of Caesar.[116]

To this, one may object that what the *divine decrees story* really shows is that, once a particular set of premises (amongst which are God's free decrees) has been assumed, then a given conclusion follows; and that on different assumptions (under different divine decrees) we have to expect different conclusions. But the question is: what does it properly mean that a property belonging to a complete concept 'follows' from a given set of assumptions? Involved here is the vexing question of the so-called *counterfactual identity*.

If the divine decrees are a constitutive part of a complete concept and if the complete concept determines the identity of the individual substance subordinate to it, then when the divine decrees change, this automatically means that the corresponding individual substance (real or possible) changes. And this, in its turn, means that if 'our' Caesar (the Caesar 'inhabiting' our world) crosses the Rubicon, it is another Caesar, not ours, who abstains from crossing it. Again, *our* Caesar seems not to have any other alternative than that of crossing the Rubicon.[117]

There are writings in which Leibniz seems to accept this consequence (foreshadowing something very similar to David Lewis's counterpart theory):

It, therefore, also follows that he would not have been our Adam, but another Adam, had other events happened to him, for nothing

---

[115] *Philosophical Essays*: 70.
[116] In this case the proposition 'Caesar crossed the Rubicon' is contingent because it is true and could have been false.
[117] On this point, see Mates (1986: 137–51); Mondadori (1973), (1975), (1985), and (1993); Cover and Hawthorne (1999).

prevents us from saying that he would be another. Therefore, he is another.[118]

You will object that you may complain that God did not give you more strength than he has. I answer: if he had done that, you would not exist, for he would have produced not you but another creature.[119]

But someone else may say, how does it come that this man will certainly commit this sin? The reply is easy; it is that otherwise he would not be this man.[120]

Many future conditionals are silly; thus, when I ask what would have happened if Peter had not denied Christ, it is asked what would have happened if Peter had not been Peter, because denying is contained in the complete notion of Peter.[121]

In the last part of the *Theodicy*, presenting his theory of possible worlds, Leibniz alludes to a world different from ours, where the Roman Sextus Tarquinius, instead of raping Lucretia, chooses to travel to Corinth, where he becomes a rich and respectable man. In this case, Leibniz explicitly assumes that the man of whom he is speaking *is not our Sextus*, but someone *similar* to him:

These worlds are all here, that is, in ideas. I will show you some, wherein shall be found, not absolutely the same Sextus as you have seen (that is not possible; he carries with him always that which he shall be), but several Sextuses resembling him, possessing all that you know already of the true Sextus, but not all that is already in him imperceptibly, nor in consequence all that shall yet happen to him. You will find in one world a very happy and noble Sextus, in another a Sextus content with a mediocre state, a Sextus, indeed, of every kind and endless diversity of forms.[122]

Moreover, quite consistently, Leibniz thinks that the world in which Sextus is placed in a city *similar* to Rome, is a world different from the actual one:

---

[118] *Philosophical Essays*: 73.   [119] A VI, 4B: 1639.
[120] L: 322.   [121] Grua, vol. 1: 358.   [122] T: 376.

If Jupiter had placed here a Sextus happy at Corinth or king in Thrace, it would no longer be this world.... They passed into another hall, and lo!, another world, another Sextus, who, issuing from the temple, and having resolved to obey Jupiter, goes to Thrace.[123]

There are other texts, however, in which Leibniz seems clearly uneasy with abandoning counterfactual identity. A clear example of this is offered by a remark on Ludovico de Dola's book (1683–5), in which Leibniz discusses the biblical episode of David taking refuge in the town of Keila:

In my opinion, middle science reduces itself to science of simple intelligence, i.e. to the science of possibles. In other words, there is a science concerning whether the inhabitants of the town of Keila, once besieged, will surrender the town to Saul, and another science which does not concern these inhabitants of Keila, whose complete concept implies that they will not be besieged, but some other possible inhabitants of the town, who have everything in common with these, except for what is coherent with the hypothesis of being besieged.[124]

Here, Leibniz at first plainly denies counterfactual identity. The inhabitants of Keila who are not besieged are not the same inhabitants who are besieged, because their complete concepts are different. Leibniz, however, seems not to be fully persuaded of this conclusion, and in the manuscript, near to the words 'another science which does not concern these inhabitants of Keila, whose complete concept implies that they will not be besieged', he puts the following marginal note:

Instead, it does concern precisely these inhabitants, because their notion only contingently involves [the property of] not being besieged.[125]

On this occasion, Leibniz seems to claim that contingent properties do not contribute to determine the identity of an individual substance, a

[123] T: 377.   [124] A VI, 4B: 1789–90.   [125] Ibid.

claim which is quite at odds with what he maintains until the end of his life, i.e. that 'to a complete concept nothing inheres by accident.'[126]

3.3 Solution 2

To introduce this solution, let me first emphasize a point about Leibniz's idea of a *demonstration*. According to Leibniz, a demonstration is a sequence of sentences (propositions, strings of symbols expressing propositions) *of a finite length*. In a paper on freedom, written two or three years after the *GI*, he gives a very simple example of demonstration, based on the principle of substitutivity:

> For demonstrating is nothing but displaying a certain equality or coincidence of the predicate with the subject (in the case of a reciprocal proposition) by resolving the terms of a proposition and substituting a definition or part of one for that which is defined, or in other cases at least displaying the inclusion so that what lies hidden in the proposition and was contained in it virtually is made evident and explicit through demonstration. For example, if by a ternary, senary, and duodenary number we understand one divisible by 3, 6, and 12, then we can demonstrate the proposition that every duodenary number is senary. For every duodenary number is a binary-binary-ternary (which is the resolution of a duodenary into its prime factors, $12 = 2 \times 2 \times 3$, that is, the definition of a duodenary), and every binary-binary-ternary is binary-ternary (which is an identical proposition), and every binary-ternary is senary (which is the definition of senary, $6 = 2 \times 3$). Therefore, every duodenary is senary ($12 = 2 \times 2 \times 3$, and $2 \times 2 \times 3$ is divisible by 2×3, and 2×3 is equal to 6. Therefore, 12 is divisible by 6).[127]

In the same text, Leibniz distinguishes two kinds of truths: immediate and derivative. In their turn, derivative truths divide into two kinds again: those that can be reduced to basic truths and those that, in the attempt to reduce them, give rise to an infinity of steps:

---

[126] A VI, 4A: 306.   [127] *Philosophical Essays*: 96.

Basic truths are those for which we cannot give a reason; identities or immediate truths, which affirm the same thing of itself or deny the contradictory of its contradictory, are of this sort. Derivative truths are, in turn, of two sorts, for some can be resolved into basic truths, and others, in their resolution, give rise to a series of steps that go on to infinity. The former are necessary, the latter contingent. Indeed, a necessary proposition is one whose contrary implies a contradiction.[128]

Thus, any truth that is demonstrated is *necessary*, whereas any proposition that is true but whose truth cannot be demonstrated (in a finite number of steps) is contingent. Leibniz explains the difference between these two kinds of truth as follows:

And here the mysterious distinction between necessary and contingent truths is revealed, something not easily understood unless one has some acquaintance with mathematics. For in necessary propositions, when the analysis is continued indefinitely, it arrives at an equation that is an identity; this is what it is to demonstrate a truth with geometrical rigour. But in contingent propositions one continues the analysis to infinity through reasons for reasons, so that one never has a complete demonstration, though there is always, underneath, a reason for the truth, but the reason is understood completely only by God, who alone traverses the infinite series in one stroke of mind.[129]

It is quite remarkable that here Leibniz does not say that God, who dominates the infinite, can demonstrate a contingent truth: God understands the reason or reasons that determine a contingent truth, but he too cannot demonstrate it. This point is confirmed in the paper on freedom mentioned above:

However, just as incommensurable proportions are treated in the science of geometry, and we even have proofs about infinite series, so to a

---

[128] Ibid.
[129] *Philosophical Essays*: 28 (translation slightly modified) (A VI, 4B: 1650).

much greater extent, contingent or infinite truths are subordinate to God's knowledge, and are known by him not, indeed, through demonstration (which would imply a contradiction) but through his infallible intuition [*visio*].[130]

Thus, the proposition 'Caesar crossed the Rubicon' being true, the predicate 'crossing the Rubicon' inheres in Caesar's complete concept, but this inherence cannot be demonstrated.

To find this solution, Leibniz drew inspiration from the Euclidean algorithm for finding the greatest common divisor or for comparing two lines of different length. The algorithm, in the case of two segments *a* and *b*, with *a* greater than *b*, operates as follows. First of all, let us subtract *b* from *a* as many times as possible. Suppose that *b* is contained in *a* 2 times with a segment *c*, shorter than *b* as a remainder. Then, subtract *c* from *b* and if there is a remainder again, call it *d*, subtract it from *c*. Suppose now that the new remainder *e* is contained exactly 2 times in *d*, without any remainder. Then *e* is the greatest segment that divides exactly *a* and *b*, and these are commensurable. There are other cases, however, in which the procedure 'can be continued to infinity, as happens in the comparison between a rational number and an irrational number, such as the comparison of the side and the diagonal of a square'.[131] Leibniz sees a strong analogy between the Euclidean algorithm and logical demonstration:

So, similarly, truths are sometimes provable, that is, necessary, and sometimes they are free or contingent, and so cannot be reduced by any analysis to an identity, to a common measure, as it were. And this is an essential distinction, both for proportions and for truths.[132]

Thus, it is on this analogy that Leibniz's second solution of the problem of contingency rests.

Leibniz was very fond of this solution, which he probably considered *the* solution of the problem of contingency. At the same time, however, he was quite reticent about it: with the exception of a cryptic hint in one

[130] *Philosophical Essays*: 97. [131] Ibid. [132] Ibid.

appendix to the *Essays of Theodicy*, indeed, he never mentioned it in his published works.[133] This attitude was certainly due to the fact that, as Leibniz himself recognized, to understand this solution, some acquaintance with mathematics was needed.

Now, a reason for carefully reading the *GI* is that it is there that for the first time Leibniz, in §§132–7, works out the analogy between logical analysis and the Euclidean algorithm. At this stage, however, he seems to be tempted to claim that contingent truths can be demonstrated, even though they cannot be reduced to identical propositions:

(132) Every true proposition can be proved, for since, as Aristotle says, the predicate is in the subject or the notion of the predicate is involved in the notion of the subject understood completely, it must be possible for the truth to be shown, at any rate by means of a resolution of the terms into their values, i.e. of those terms that they contain.[134]

(134) A true contingent proposition cannot be reduced to identical propositions: it is proved, however, by showing that if the resolution is continued further and further, it incessantly approaches the identical propositions, but it never reaches them. Therefore, only God, who embraces the whole infinite in his mind, has the power of knowing all contingent truths with certainty.[135]

What drives Leibniz to this conclusion is again the mathematical analogy. Since it is possible to prove that 'an asymptote, for instance, incessantly approaches another and that two quantities (also in the case of asymptotes) are equal each other, by showing what will be the case, once the progression is continued as far as we please,'[136] Leibniz thinks that we human beings will also be able to gain certitude of contingent truths by showing that 'the progression is continued as far we please'. But he suddenly realizes that in the case of truths the analogy with mathematics can be developed until to a certain point only:

But we must reply that there is surely a similitude here, but not an agreement in all respects. And that there can be respects which, by an analysis continued as far as we please, will never reveal themselves

---

[133] Cf. GP 6: 414.   [134] *GI*: 21 (*A* VI, 4A: 776).   [135] Ibid.   [136] Ibid.

with a sufficient degree of certainty, and which are seen perfectly only by him whose intellect is infinite. Of course, as with asymptotes and incommensurables, so with contingent things we may observe many things with certainty, from the very principle that every truth must be provable. Consequently, if all things are alike on one side and on the other of our hypotheses, then there can be no difference in the conclusions. Analogously, we may observe other things of this sort, which are true in the case of necessary and of contingent propositions as well, because they are reflexive. But we can no more give the full reason of contingent things than we can forever follow the asymptotes and go along the infinite progressions of numbers.[137]

Oddly enough, Leibniz never attempts to integrate the solution of the divine decrees and that of the infinite analysis into a single coherent theory. He employs each of them on different occasions in different texts and seems not to be interested in explaining what kind of relationship subsists between them.

To the best of my knowledge, however, there is at least one text in which Leibniz seems to think of a solution based neither on the divine decrees nor on the infinite analysis: it is a text written ten years after the *Discourse on Metaphysics* in which Leibniz develops some remarks on a work of the theologian William Twisse devoted to the so-called *middle knowledge*. This is the relevant passage from Leibniz's remarks:

> when I ask what would have happened if Peter had not denied Christ, it is asked what would have happened if Peter had not been Peter, for denying is contained in the complete notion of Peter. But it is justifiable that by the name Peter should be understood those attributes that inhere in him, from which the denial does not follow, while at the same time there must be subtracted from the whole universe everything from which it does follow: then, sometimes it can happen that the decision follows per se from the remaining things posited in the universe, but sometimes it will not follow unless with the addition of a new divine decree *ex ratione optimitatis.*[138]

---

[137] Ibid.
[138] Grua, vol. 1: 358 (the original has been slightly modified according to the manuscript).

In this case, Leibniz emphasizes that the complete concepts correspond-ing to, respectively, *Peter the Apostle who denied Christ* and *Peter the Apostle who did not deny Christ* share a large number of properties that authorize us to employ a single name to denote both. Obviously, if we assume that similarity between two or more individuals is based on the fact of sharing one or more properties, the remark about Twisse agrees with what Leibniz writes in the last chapters of the *Theodicy*.[139] In this passage, however, there is something more than an obvious remark con-cerning the similarity of two individuals and the use of proper nouns. Leibniz here hints at the relationship between a given property belong-ing to a complete concept and other properties from which it derives and that are internal and external to the complete concept. In the pas-sage in question, Leibniz argues as follows.

If the individual the Apostle Peter belonging to our world denies Christ, then the property of *denying Christ* must inhere in the complete concept corresponding to him. The denial, in this case, follows from properties inhering in Peter's complete concept and from some proper-ties 'external' to it. If we 'subtract' from Peter's complete concept and the possible world to which Peter belongs all the properties from which the denial follows, then it is still possible to employ the noun 'Peter' to denote what remains of Peter's complete concept, once the causes of the denial have been removed from it.

Acknowledgments

I am deeply indebted to Richard Arthur, Wolfgang Lenzen and Marko Malink, who have much improved the first draft of this work with com-ments and suggestions. Gian Biagio Conte, and occasionally Gianpiero Rosati, have helped me with some difficult passages in Leibniz's Latin. Maria Rosa Antognazza, Dan Garber and Paolo Mancosu constantly encouraged me to publish this book: to all of them I express my gratitude.

---

[139] On Leibniz and counterfactual identity, see, in particular, Adams (1994: 52–74). On Leibniz, infinite analysis and possible worlds, see Bernini (2002).

Generales Inquisitiones de Analysi Notionum et Veritatum.

1686[1]

Omittamus nunc quidem omnia Abstracta, ita ut Termini quicunque non nisi de Concretis, sive ea sint substantiae, ut *Ego*, sive phaenomena, ut *Iris*, intelligantur. Itaque nec de discrimine inter abstracta et concreta nunc erimus soliciti, vel saltem non alia nunc adhibebimus abstracta, quam quae sunt Logica seu Notionalia, verb. grat. ut *Beitas* ipsius A, nihil aliud significet quam τὸ A esse B.

Privativum non-A.

Non-non-A idem est quod A.

Positivum est A, si scilicet non sit non-Y quodcunque, posito Y similiter non esse non-Z et ita porro.

Omnis terminus intelligitur positivus, nisi admoneatur eum esse privativum.

Positivum idem est quod Ens.

Non Ens est quod est mere privativum, seu omnium privativum, sive non-Y, hoc est non-A, non-B, non-C, etc. Idque est quod vulgo dicunt nihili nullas esse proprietates.

Omnem quoque Terminum hic accipiemus pro completo, seu substantivo, ita ut *magnus* idem sit quod *Ens magnum* sive ut ita dicam *magnio*, quemadmodum qui nasutus est dicitur *naso*, itaque in his adjectivi et substantivi discrimine non indigemus, nisi forte ad emphasin significandi.

Ens est vel per se vel per accidens, seu terminus est necessarius vel mutabilis. Ita *Homo* est Ens per se; at Homo doctus, Rex, sunt Entia per

---

[1] Hic egregie progressus sum.

General Inquiries about the Analysis of Notions and Truths.

1686 [a]

Let us omit for the present at least, all abstract things, so that any kind of terms are understood to refer to concrete things alone, whether they be substances, like the *I*, or phenomena, like the *Rainbow*.* Thus, we shall not be concerned now with the distinction between abstract and concrete things, or, at any rate we shall not introduce any kind of abstract things other than those which are logical or which concern the notions, so that, for instance the *B-ness* of A does not mean anything else than *the being B of A*.*

Non-A is *Privative*.*

*Non-non-*A is the same as A.

A is *Positive*, if it is not the case that it is not any non-Y whatsoever, it being assumed that Y is not non-Z, and so on.

Every term is understood to be positive, unless notice is given that it is privative.

'Positive' is the same as 'being'.*

Non-being is that which is merely privative, i.e. privative of everything, or non-Y, i.e. non-A, non-B, non-C, etc. This is what is meant when it is commonly said that the nothing has no properties.*

Moreover, we shall consider here every term as complete, i.e. as a substantive, so that *big* is the same as *a big being* or, so to speak, *a big one*, just as someone with a big nose is called *nose*. Thus, we do not need here the distinction between adjective and substantive, except perhaps for indicating emphasis.*

A being is either in itself or by accident, i.e. a term is either necessary or changeable. Thus, 'man' is a being in itself, whereas 'learned man', 'king'

---

* Here I have made excellent progress.

accidens. Nam res illa quae dicitur Homo, non potest desinere esse homo, quin annihiletur; at potest quis incipere vel desinere esse Rex aut doctus licet ipse maneat idem.

Terminus est vel *integralis* sive perfectus, ut Ens, ut Doctus, ⟨ut idem vel similis ipsi A, qui scilicet potest esse subjectum vel praedicatum propositionis, licet nihil accedat⟩ [;] vel est *partialis* sive imperfectus, ut: idem, similis; ubi aliquid addendum est (nempe: ipsi A) ut integer terminus exurgat. Et vero id quod addendum est, oblique accedit; rectum enim integrali accedens salva termini integritate semper addi et omitti potest. Et *in recto* junguntur duo termini integrales constituentes novum integralem. Interim non omnis terminus cui alius *in obliquo* additur partialis est; ita Ensis est integralis, licet oblique addendo inde fiat Ensis Evandri. Itaque potest aliquid nonrectum salva integritate termini omitti, ut hoc loco obliquum: Evandri. At contra obliquum recto omisso integralem terminum non facit. Et proinde si terminus per se integralis cum aliqua flexione vel connexionis nota alteri addatur ita ut altero omisso integralem non faciat, additus est in obliquo. Potest autem ex obliquo a recto divulso fieri integralis, ut ex obliquo: Evandri, fieri potest qui est res Evandri, seu Evandrius. Utile autem erit curare ut termini integrentur. Et proinde opus erit signis quibusdam rerum vel terminorum generalibus; ita si volumus semper uti in nostra characteristica non nisi terminis integralibus, non dicendum erit Caesar est similis Alexandro, sed Caesar est similis $\tau\hat{\omega}$ A qui est Alexander ⟨seu similis rei quae est Alexander⟩. Itaque terminus noster non erit: similis; sed: similis $\tau\hat{\omega}$ A. Eodem modo non exprimetur verbotenus: Ensis Evandri, sed Ensis qui est res Evandri, ita ut: qui est res Evandri, sit unus terminus integralis. Hoc modo poterimus dividere terminum quemlibet compositum in integrales. Sed haec quousque et qua ratione exequi liceat, progressus docebit. ⟨Quod si hoc semper procedit, non alia habebimus nomina quam integralia. Videbimus an *ex ipsis particulis* similiter integralia formare liceat, ut pro A in B, A *inexistens in aliquo*, quod est B.⟩

are accidental beings.* For that thing which is called 'man' cannot cease to be a man without being annihilated; but someone can begin or cease to be a king, or learned, while continuing to remain the same. A term is either *integral*, i.e. perfect, like, for instance, 'being', 'learned', ⟨'identical' or 'similar to A', if it can be the subject or predicate of a proposition without any addition⟩ [;] or it is *partial*, i.e. imperfect, like for instance: 'identical', 'similar', where something has to be added (namely, 'to A') for an integral term to arise. And what has to be added, indeed, is added obliquely; a direct term added to an integral one can always be added and omitted without affecting the integrity of the term.* And two integral terms joined *directly* constitute a new integral term. Not every term, however, to which another is added *obliquely* is partial; thus, 'sword' is integral, even though by oblique addition it gives rise to 'the sword of Evander'. Thus, something non-direct can be omitted without affecting the integrity of a term, as in this case the oblique term 'of Evander'. Contrariwise, an oblique term, if the direct term is omitted, does not make an integral term. Therefore, if a term which is integral in itself is added to another by means of some inflexion or mark of connection in such a way that, when the other is omitted, it does not make an integral term, then it is added obliquely. It is possible, however, to obtain an integral term when an oblique term is separated from an integral one; thus, from the oblique 'of Evander' it is possible to obtain 'that which is a thing of Evander', or 'Evandrian'. It will be useful, however, to take care that terms are made integral. And for this reason, some general signs of things or terms will be needed; thus, if we wish always to employ only integral terms in our *characteristic*,* we must say not 'Caesar is similar to Alexander', but 'Caesar is similar to A, which is Alexander' ⟨or 'similar to the thing which is Alexander'⟩. So our term will be, not 'similar', but 'similar to A'. In the same way, instead of employing literally the expression 'the sword of Evander', we should say 'the sword, which is a thing of Evander', so that 'which is a thing of Evander' is one integral term. In this way we shall be able to divide any composite term into integral ones. The progress of this work, however, will show how far and in what way this may be carried out;* ⟨and if we always succeed in this, we shall have no other nouns but integrals. We shall see whether it will similarly be possible to form integral nouns *from particles themselves*, as, for instance, 'A *existing in another*, which is B' in place of 'A in B'.⟩

Ex his patet porro esse integrales qui in partiales resolvantur, et esse rectos in quos (si resolvas seu definitionem pro definito substituas) manifestum sit ingredi obliquos. Partiales ergo, itemque particulae quae obliquis additae inde faciunt rectos, et partialibus additae faciunt integrales, prius explicari debent quam integrales, qui in partiales et particulas resolvuntur. Sed tamen ante partiales et particulas explicari debent illi integrales qui aut non resolvuntur, aut non nisi in integros. Et tales integrales a partialibus independentes utique esse necesse est, saltem generales, ut Terminus, Ens, nam his ipsi partiales indigent, ut transeant in integrales, ultimum enim complementum partialis ⟨vel obliqui⟩, ut in integralem transeat, cum sit integrale, rursus in integralem et partialem resolvi non potest. Talium integralium in obliquos et partiales nobis irresolubilium enumeratione opus est, quam reliquorum Analysis dabit, et initio satis erit eos enumerare tanquam pure integrales, quorum resolutione in non-integrales minus opus videtur. Res etiam eo reducenda est, ut paucis adhibitis integralibus per partiales et obliquos compositis, reliqui omnes inde recta seu similariter sive sine obliquis componi possint. Et ita constitui poterunt pauci integrales, vel sane certi definiti, aut definita serie progredientes, qui poterunt considerari tanquam primitivi in recta resolutione ex quibus reliqui magis compositi deinde oriantur, ut numeri derivativi ex primitivis. Eaque ratione cuilibet Notioni ⟨quatenus sine obliquitate resolvitur⟩ suus posset numerus characteristicus assignari.

Habemus igitur *1mo Terminos*[2] *integrales primitivos simplices* irresolubiles, vel pro irresolubilibus assumtos, *ut: A.*

2do Particulas simplices seu syncategoremata primitiva, *ut: in.*

---

[2] Terminum intelligo integralem, nam partiales fiunt ex integrali et particula, ut: pars, est Ens in aliquo, etc.

From this it is also clear that there are integral terms which are resolved into partial ones, and that there are direct terms into which (if you resolve them, i.e. substitute a definition for what is defined) it is manifest that oblique terms enter. Therefore, partial terms—and likewise particles which, when added to oblique terms, make direct terms out of them and, when added to partial terms, make integral terms—ought to be explained before the integrals that are analysed into partial terms and particles.* Before partials terms and particles, however, one has to explain those integral terms which either cannot be analysed or are analysed only into integral terms. And it is absolutely necessary that such integral terms, independent of the partial terms, exist, at least those that are general, as 'term', 'Being'. For even the partial terms need these integral terms to become integral, since the ultimate complement of a partial ⟨or of an oblique term⟩, which makes it into an integral one, being itself integral, cannot again be analysed into an integral and a partial term. We need an enumeration of such integral terms as cannot be analysed into oblique and partial terms, and this will be done through a resolution of all remaining terms. Initially it will be sufficient to enumerate as purely integral those terms whose resolution into non-integral seems to be less necessary. The matter is thus to be reduced to this, that, given a small number of integral terms composed of partial and oblique terms, all the remaining terms can be composed from them directly, i.e. similarly or without oblique terms. Thus, it will be possible to establish a few integral terms, either very accurately defined, or terms progressing in a definite series, which we will be able to consider as primitive terms in a direct resolution from which all the remaining more composite ones arise, as derivative numbers arise from prime numbers. With this method one can assign to each concept, ⟨in so far as it is analysed without obliquity⟩, its own characteristic number.*

We have, therefore: *(1st) simple primitive integral terms*, which are not resoluble, or assumed to be not analysable, *such as:* 'A'.[b]

(2nd) Simple particles, i.e. primitive syncategorematic terms, *such as:* '*in*'.*

---

[b] I mean an integral term, because partial terms are made from an integral term and a particle, as for example 'part' is 'a being in something else'.

3^{tio} Terminos integrales primitivos compositos ex meris Terminis simplicibus, idque recta seu sine interventu particularum vel syncategorematum, *ut: AB.*

4to Particulas compositas ex meris particulis simplicibus, sine Termini (categorematici) interventu, *ut: cum-in,* qua particula uti possemus ad designandam (si categorematicis ⟨postea⟩ adjiciatur) rem quae *cum* aliquo est *in* aliquo.[3]

5to habemus *Terminos integrales derivativos simplices.* Appello autem *derivativos,* qui oriuntur non per solam compositionem, scilicet similarem, seu recti cum recto, ut: AB, sed per flexionis cujusdam aut particulae sive syncategorematici interventum, *ut: A in B;* ubi A et B dissimilariter terminum compositum ex ipsis, nempe τὸ A in B, ingrediuntur.[4] Quam differentiam inter compositionem et derivationem quodammodo et Grammatici observant. Sunt ergo derivativi simplices, qui non possunt resolvi in alios derivativos, sed non nisi in primitivos simplices, cum particulis.

6to habemus *Terminos integrales derivativos compositos,* qui scilicet *recta* seu *similariter* componuntur ex aliis derivativis, et hi oblique etiam componuntur ex primitivis compositis una cum particulis.

7mo Ambigi potest de illis *derivativis qui constant ex primitivis simplicibus, et particulis compositis,* utrum sint potius simplices quam compositi, sane in alios categorematicos resolvi non possunt, nisi primitivi unius duplicatione, quatenus componendo eum nunc cum una nunc cum alia particula ⟨simplice, compositam componente⟩, duo novi inde fieri possunt derivativi simplices, ex quibus fieri potest propositus derivativus, quasi compositus.

---

[3] Dubito an procedat particula cum-in. Sit A cum B in C; non commode dicitur A esse cum-in BC nisi velimus per ordinem exprimi cum pertinere ad B, et in ad C.

[4] In formandis compositis non refert quo ordine collocentur simplicia, at in formandis derivativis interesse potest quo *ordine.*
Recta compositio idem quod similaris.

(3rd) Composite primitive integral terms, composed of mere simple Terms and in a direct way, i.e. without any recourse to particles or syncategorematic terms, *such as:* 'AB'.

(4th) Composite particles, composed of mere simple particles, without any recourse to a (categorematic) term, *such as:* 'with-in'. We may use this particle, once it is ⟨later⟩ added to the categorematic terms, to designate a thing that is *with* something *in* something else.[c]

(5th) We have *simple derivative integral terms.* I call *derivative* those terms which arise, not through composition alone, i.e. through similar composition of a direct term with a direct one, such as 'AB', but through the mediation of some inflexion, or of a particle or syncategorematic term, *such as* 'A *in* B', where 'A' and 'B' enter dissimilarly a term composed of them, namely 'A in B'.[d] This difference between composition and derivation is in a certain sense observed by grammarians, as well. There are, therefore, simple derivative terms, which cannot be analysed into other derivative terms, but only into simple primitive terms with particles.

(6th) We have *composite derivative integral terms,* namely those which are compounded *directly,* i.e. *similarly,* from other derivative terms, and these are also obliquely compounded from primitive composite terms together with particles.

(7th) Regarding those *derivative terms that consist of primitive simple terms and composite particles,* we may doubt whether they are simple rather than composite. Certainly, they cannot be analysed into other categorematic terms, except by duplication of one primitive term, in so far as, by compounding this now with one and now with another ⟨simple particle, forming a composite⟩, two new simple derivative terms can be made, from which the proposed derivative term may arise, as if it were composite.

---

[c] I wonder if the particle 'with-in' does really work. Suppose that A with B is in C: it does not fit to say 'A is with-in BC' unless we want to mean that there is an *order* according to which 'with' concerns B and 'in' concerns C.

[d] When we are forming composite terms, the order according to which the simple ones are disposed does not matter, whereas for forming the derived terms *the order* according to which they are disposed may be of consequence.

The direct composition is the same as the similar one.

Octavo quemadmodum habemus Terminos categorematicos primitivos et derivativos, ita et haberi possunt *particulae derivativae*, eaeque rursus *simplices* quidem ex particula simplice et termino primitivo; At [compositae][5] (*Nono*) ex particula composita et termino primitivo, quae resolvi possunt in plures particulas derivativas simplices.

Et *decimo* hic similiter ambigitur quid dicendum de particula derivativa ex termino primitivo composito et particula simplice. ⟨Fortasse tamen praestat efficere ut omnes particulae cessent, quemadmodum et omnes obliqui, quemadmodum pagina praecedente dictum est. Nisi tamen obstet, quod ita non facile apparebit, quae quibus sint arreferenda.⟩

Illud tamen adhuc in considerationem venire debet quod particulae etiam primitivae simplices non uniuntur ita similariter, ut termini primitivi simplices. Itaque multae in compositione particularum occurrere possunt varietates. Exempli causa, si dicam Johannes-Pauli-Petri, id est Johannes Petri, qui fuit Pauli est quaedam compositio similaris; at si dicam Socrates Sophronisci ex Athenis, dissimilaris est particularum vel flexionum compositio. Et hinc orientur haud dubie varii respectus, variaeque obliquitates obliquitatumque mixturae, quarum accurata constitutione potissima characteristicae artis pars continetur. Sed de his non satis potest judicari antequam primitiva simplicia tam in Terminis quam in particulis prorsus accurate constituantur; vel saltem pro illis interim derivativa quidem et composita, sed primitivis simplicibus propiora assumantur, donec paulatim ad ulteriorem resolutionem via se sponte aperiat. Sub particulis etiam hoc loco comprehendo nomina primitiva partialia, si qua sunt quae in alia primitiva partialia non possint resolvi. Sed revera puto ea fieri ex Ente vel alio integrali termino cum particula.

Termini primitivi Simplices vel interim pro ipsis assumendi, sunto:

Terminus (quo comprehendo tam Ens quam Non-Ens). *Ens* seu possibile (intelligo autem semper concretum quia abstracta tanquam non necessaria exclusi). *Existens* (licet revera reddi possit causa existentiae, et definiri posset Existens, quod cum pluribus compatibile est, quam

---

[5] [L: compositas].

(8th) Just as we have primitive and derivative categorematic terms, so we may also have *derivative particles*. And these again are either *simple*, if composed of a simple particle and a primitive term; or

(9th) composites of a composite particle and a primitive term. These can be analysed into several derivative simple particles.

(10th) We may similarly doubt here what is to be said of a derivative particle composed of a primitive composite term and a simple particle. ⟨Perhaps, however, it is more convenient to arrange things so that all particles together with all oblique terms are left aside, just as we said on the previous page—unless it be objected that in this way it will not be readily apparent which are related to which.*⟩

Now we have to consider, however, that even simple primitive particles are not joined in a similar way to that in which simple primitive terms are. Thus, many varieties can occur in the composition of particles. For example, if I say 'John-of-Paul's-Peter', i.e. 'John, the son of Peter, who was the son of Paul', there is a kind of similar composition; whereas, if I say 'Socrates of Sophroniscus, from Athens', the composition of particles and flexions is dissimilar. Undoubtedly from this there will arise various respects and various obliquities and mixtures of obliquities, the accurate constitution of which comprises the chief part of the characteristic art. But no satisfactory judgement can be made concerning these before the simple primitives, both in terms and in particles, are first accurately established; or, at any rate, before we temporarily assume in their place terms that are indeed derivative and composite, but closer to the simple primitive simple ones, until gradually the way to a further resolution spontaneously opens by itself. Under the particles I also include here some partial primitive nouns, if there are any which cannot be analysed into other primitive partials. But, as a matter of fact, I believe that they arise from 'being', or another integral term, with a particle.

Let the *simple primitive terms*, or those to be assumed temporarily in their place, be the following:

'*Term*' (by which I understand both 'being' and 'non-being').* '*Being*' or what is possible (I always understand a concrete thing, because I have excluded abstract things as unnecessary). '*Existent*' (even though, in fact, it is possible to assign a cause to existence, and 'existent' can be defined as

quodlibet aliud incompatibile cum ipso. Nos tamen his tanquam altioribus nunc abstinemus). *Individuum* (etsi enim Ens omne revera sit individuum, nos tamen terminos definimus, qui designant, vel quodlibet individuum datae cujusdam naturae, vel certum aliquod in[di]viduum determinatum, ut Homo seu quilibet homo, significat quodlibet individuum naturae humanae particeps. At certum individuum est Hic; quem designo vel monstrando vel addendo notas distinguentes, quanquam enim perfecte distinguentes ab omni alio individuo possibili haberi non possint, habentur tamen notae distinguentes ab aliis individuis occurrentibus). *Ego* (est aliquid peculiare, et difficulter explicabile in hac notione, ideo cum integralis sit, ponendam hic putavi).

Sunt etiam Termini primitivi simplices omnia illa phaenomena confusa sensuum, quae clare quidem percipimus, explicare autem distincte non possumus, nec definire per alias notiones, nec designare verbis. Ita caeco quidem multa dicere possumus de extensione, intensione, figura, aliisque varietatibus quae colores comitantur, sed praeter notiones distinctas comites est aliquid in colore confusum, quod caecus nullis verbis nostris adjutus concipere potest, nisi ipsi aliquando oculos aperire detur. Et hoc sensu, album, rubrum, flavum, caeruleum, quatenus in illa inexplicabili imaginationis nostrae expressione consistunt, sunt termini quidam primitivi. Utile tamen erit eos cum confusi sint, ratiocinationemque nihil adjuvent, evitare quoad licet, adhibendo loco definitionum notiones distinctas comites, quatenus eae sufficiunt ad confusas inter se discernendas. Interdum et miscere ambas methodos inter se, utile erit, prout commoditas dabit. Itaque primariis istis proprias notas dare possumus, caeteris per eas explicatis. Sic coloratum est terminus explicabilis per relationem ad nostros oculos, sed quia ea relatio sine multis verbis accurate exprimi non potest, et ipse oculus rursus explicatione prolixa indigeret, tanquam machina quaedam, poterit coloratum assumi ut terminus primitivus simplex, cui addendo notas quasdam differentiales, poterunt designari colores varii. Fortasse tamen coloratum

that which is compatible with more things than anything else incompatible with it. At present, however, we abstain from these, as it were, loftier considerations). 'Individual' (even though every being is really an individual, we are defining terms which designate either any individual whatsoever of a certain given nature, or some certain determinate individual. Thus, for instance, 'Man' or 'any man' means any individual who participates in human nature. A particular individual, however, is *this* one, whom I designate either by pointing out or by adding some distinguishing signs. For, although it is impossible that there are marks which distinguish it perfectly from any other possible individual, there are however marks such that distinguish it from other individuals that we meet). 'I' (there is something peculiar and difficult to explain in this notion, but since it is integral, I thought it should be put here).*

Also, simple primitive terms are all those confused phenomena of the senses, that we perceive clearly but which, however, we are not able to explain distinctly, or define by means of other notions, or designate by words. Thus, there is no doubt that we can say much to a blind man about extension, intensity, shape, and various other properties which accompany colours; but besides these accompanying distinct notions, there is something confused in colour which a blind man cannot conceive with the help of any words of ours, unless it is granted to him to open his eyes at some time. In this sense, 'white', 'red', 'yellow', 'sky-blue', in so far as they consist of that inexplicable expression of our imagination, are primitive terms of a kind. But, because they are confused and are of no help for our reasoning, it will be useful to avoid them as far as possible by employing instead of definitions the distinct notions, which accompany them, as far as these are sufficient to distinguish between the confused notions. Sometimes it will even be useful to mix both methods as convenience dictates. Therefore, to these primary terms we may give their marks, all the remaining terms being explained by means of these. Thus 'coloured' is a term which may be explained through its relation to our eyes; but because that relation cannot be expressed accurately without many words, and the eye itself, as a kind of machine, again needs a lengthy explanation, it will be possible to take *coloured* as a primitive simple term, and adding to it certain differentiating marks the various colours can be designated. Perhaps, however, 'coloured' can be defined

definiri poterit per perceptionem superficiei sine sensibili contactu. Sed horum quid praestet in progressu patebit.

Videntur inter primitivos simplices recenseri posse omnes notiones quae continent materiam cujusdam quantitatis, sive in quibus res homogeneae conveniunt inter se, ut habens magnitudinem; extensum, durans, intensum, sed hae notiones ni fallor resolvi adhuc possunt. De Notionibus Extensi et cogitantis peculiariter dubitari potest an sint simplices; multi enim arbitrantur has esse notiones quae per se concipiantur, nec porro resolutione indigeant, sed Extensum videtur esse continuum habens partes coexistentes. Et terminus cogitantis videtur non esse integralis, refertur enim ad aliquod objectum quod cogitatur. Inest tamen in ipsa cogitatione realitas aliqua absoluta quae difficulter verbis explicatur. Et in extensione videmur aliquid aliud concipere, quam continuitatem et existentiam. Nihilominus satis videtur plena notio extensionis, ut concipiamus coexistentiam continuatam, sic ut omnia coexistentia faciant unum, et quodlibet in extenso existens sit continuabile seu repetibile continue. Interea si e re videretur Extensum, vel etiam situm (seu in spatio existens), assumere ut primitiva simplicia, ut et cogitans (seu Unum plura exprimens cum actione immanente, seu conscium) nihil ea res noceret, si praesertim deinde adjiciamus axiomata quaedam unde caeterae omnes propositiones adjunctis definitionibus deducantur. Sed haec omnia, ut saepe dixi, ex ipso progressu melius apparebunt. Et praestat progredi, quam nimia quadam morositate obhaerescere in ipsis initiis.

Tentemus nunc explicare *Terminos partiales* seu respectivos ex quibus et particulae nascuntur, notantes respectum terminorum. Quod primum mihi inquirenti occurrit est *idem*. Idem autem esse A ipsi B, significat alterum alteri substitui posse in propositione quacunque salva veritate.[6] Nam respectus illi per propositiones sive veritates explicantur. Sic

---

[6] Videndum an posito A ubique substitui posse ipsi B, etiam vicissim sequatur B ubique posse substitui ipsi A, sane si termini isti se similiter habent in relatione inter se invicem, utique mutua est substitutio. Quodsi non se habent similiter, nec ad tertium quodlibet plane eodem modo se habent, nec proinde unum alteri substitui poterit.

through the perception of a surface without sensible contact. But which of these solutions is the best will become clear as we progress.*

It seems that among the primitive simple terms we may number all notions containing matter of a certain quantity, i.e. in which homogeneous things agree, as for instance 'having magnitude'; 'extended', 'enduring', 'intensive', but these notions, unless I am mistaken, can undergo a further resolution. In particular, it can be doubted whether the Notions of 'Extended' and 'thinking' are simple; for many think that these are notions, which are conceived in themselves and need no further resolution.* 'The extended', however, seems to be something continuous having coexisting parts, and the term 'thinking' seems not to be integral, since it refers to some object which is thought. In thought itself, however, there is some absolute reality that is difficult to explain in words; and in extension we seem to conceive something else besides continuity and existence. Nevertheless, the notion of extension seems to be complete enough for us to conceive a continued coexistence, such that all coexistents make one, and whatever exists in the extended thing is continuable, i.e. continuously repeatable. Meanwhile, if it seems to the point to assume as simple primitive terms 'extended', or even 'situation' (i.e. existing in space), and 'thinking' as well (i.e. one thing expressing a plurality of things with immanent action or conscious), there is no harm done, especially if we add certain axioms from which all the remaining propositions are deduced, once the definitions are added. But all this, as I have often said, will become clearer as we proceed further. And it is better to proceed, than to remain stuck at the very beginning because of a fastidious pedantry.

Let us now try to explain *partial*, i.e. respective *Terms*, from which there also arise the particles denoting the relation of terms. What comes to mind first in my inquiry is *the same*. That A is the same as B means that one can be substituted for the other in any proposition whatsoever without loss of truth.ᵉ For those respects are explained by means of

---

ᵉ We must see whether, once it is assumed that A can everywhere be substituted for B, it also follows that conversely B can be substituted everywhere for A; certainly, if these terms behave similarly in their inter-relations, then the substitution is mutual. But if they do not behave similarly, then neither are they related exactly in the same way to a third term, nor consequently can the one be substituted for the other.

Alexander Magnus, et rex Macedoniae victor Darii; item triangulum, et trilaterum, sibi substitui possunt. Porro haec coincidere ostendi semper potest resolutione, si scilicet eo usque resolvantur, donec appareat a priori esse ipsa possibilia, si etiam formaliter prodeant iidem termini, tunc diversi termini sunt iidem. Sit Terminus A et terminus B, si pro utroque substituatur definitio, et pro quolibet ingrediente rursus definitio, donec perveniatur ad primitivos simplices, prodibit in uno, quod in alio seu formaliter idem, Ergo A et B erunt *coincidentes*, seu iidem virtualiter. Sic ergo definiri potest:

Coincidit A ipsi B, si alterum in alterius locum substitui potest salva veritate, seu si resolvendo utrumque per substitutionem valorum ⟨(seu definitionum)⟩ in locum terminorum, utrobique prodeunt eadem, eadem inquam formaliter, ut si utrobique prodeat L, M, N. Salva enim veritate fiunt mutationes quae fiunt substituendo definitionem in locum definiti vel contra. Hinc sequitur si A coincidit ipsi B etiam B coincidit ipsi A.

Proxima notio, ut A sit *subjectum,* B *praedicatum*, si B substitui potest in locum ipsius A salva veritate, seu si resolvendo A et B, eadem quae prodeunt in B prodeunt etiam in A. Idem aliter explicari potest, ut A sit B, si *omne* A, et *quoddam* B coincidant.

Habemus igitur Notas: *Coincidens* ipsi B. *Subjectum* et *praedicatum. Est. Omne. Quoddam.*

Si dicatur quoddam A est B, sensus est: quoddam A et quoddam B coincidunt. Unde et sequitur quoddam B est A.

Si omne A et quoddam B coincidunt, etiam quoddam A et quoddam B coincidunt. Sed hoc tamen videtur posse demonstrari ex negativis, ad ea igitur accedamus.

Ut A et A sunt prima coincidentia, ita A et non-A sunt prima disparata. *Disparatum* autem est, si falsum est quoddam A esse B. Itaque si B = non-A, falsum est quoddam A esse B.

Generaliter si A sit B, falsum est A esse non-B.

Si falsum sit quoddam A esse non-B, dicetur Nullum A esse non-B, seu Omne A esse B. Hinc demonstrari poterit haec consequentia: *Omne A est B. Ergo quoddam A est B.* Hoc est: Omne A et quoddam B

propositions, i.e. truths. Thus, 'Alexander the Great' and 'the king of Macedonia who conquered Darius', and likewise 'triangle' and 'trilateral', can be substituted for each other. Moreover, that these coincide may always be shown by resolution, namely if they are analysed until it appears a priori that they are possible, and if the same terms appear formally, then different terms are the same. Let there be a term A and a term B: if for each of the two their definition is substituted, and for any constituent term its definition again, until we arrive at simple primitive terms, the result in one will be what it is in the other, i.e. formally the same. Therefore, A and B will be *coincidents*, i.e. virtually the same. Thus, we may give the following definition:

'*A coincides with B*' if the one can be substituted in place of the other without loss of truth, i.e. if, on analysing each of the two by substitution of their values ⟨(i.e. of their definitions)⟩ for the terms, the same things appear on both sides: the same, I mean, formally, as, for example, if L, M, and N appear on both sides. For those changes are made without loss of truth which are made by substituting the definition for the defined term, or conversely. Hence it follows that if A coincides with B, B also coincides with A.*

The next notion is that A is the *subject*, B the *predicate*, if B can be substituted for A without loss of truth, i.e. if on analysing A and B, the same things which appear in B also appear in A. The same can be explained in another way: A is B if *every* A and *some* B coincide.

We have, therefore, the signs: *coincident* with B; *subject* and *predicate*. *Is. Every. Some.*

If one says 'some A is B', the sense is: some A and some B coincide. Hence from this, 'some B is A' also follows.

If every A and some B coincide, then some A and some B also coincide. But it seems that this can be demonstrated from negatives, so let us proceed to them.

Just as A and A are first coincidents, so A and non-A are the first disparates. Something is *disparate* if it is false that some A is B. Therefore, if B = non-A, then it is false that some A is B.*

In general, if A is B, it is false that A is non-B.

If it is false that some A is non-B, it will be said that no A is non-B, i.e. that every A is B.

coincidunt. Ergo quoddam A et quoddam B coincidunt. Nam Si Omne A et quoddam B coincidunt, Ergo falsum est quoddam A et quoddam non-B coincidere (ex definitione *Omnis*). Ergo verum est quoddam A et quoddam B coincidere.

Sed operae pretium est totam rem Enuntiationum, et respectus terminorum qui ex variis enuntiationibus nascuntur, tractare accuratius. Inde enim Origo plerorumque Terminorum partialium, et particularum sumenda est.

⟨Quaelibet litera ut A, B, L, etc. significat mihi vel terminum aliquem integralem, vel integram aliam propositionem.

Cum pro pluribus terminis ponitur unus, illi sunt definitio seu valor assumtitius, hic definitum, ut si pro AB pono C, seu cum A = BC est primitiva propositio.

Coincidunt A et B, si per substitutionem valorum assumtitiorum loco terminorum et contra, utrobique prodit idem, eadem formula.⟩

Coincidere dico enuntiationes, si una alteri substitui potest salva veritate, seu quae se reciproce inferunt.

(1) Coincidunt: Enuntiatio (directa) L et enuntiatio (reflexiva) L est *vera*. Hinc coincidunt L esse veram est vera (falsa), [et] L est vera (falsa).[7]

⊙ Coincidunt: L est vera, et: L esse falsam est *falsa*.

Coincidunt quod L esse falsam sit vera, et quod L est falsa. Hoc theorematis instar demonstrare possum, hoc modo: L esse falsam est enuntiatio quae vocetur M. Jam coincidit M est vera et M (per 1). Ergo pro M reddendo valorem, coincidunt L esse falsam est vera, et L est falsa.

Idem aliter, licet prolixius, adhibendo et ⊙ sic demonstratur: quod L esse falsam sit vera, coincidit cum hac quod L esse falsam falsam esse, est falsa (per ⊙), et ista rursus cum hac quod L esse veram est falsa (per eandem ⊙) et ista denique cum hac quod L est falsa (per 1).

---

[7] Haec potius differenda pro explicatis propositionibus.
Generaliter etsi A sit terminus, semper dici poterit A est verum, coincidit cuidam vero.

Hence this consequence can be demonstrated: *Every A is B. Therefore, some A is B*, that is: *every A* and *some B* coincide. Therefore, some A and some B coincide. For, if every A and some B coincide, then it is false that some A and some non-B coincide (by definition of *every*). Therefore, it is true that some A and some B coincide.

But it is worth our while to discuss more accurately the whole subject of statements, and of the relations of terms arising from various statements, since it is from this that the origin of many partial terms and particles is to be derived.

⟨Any letter, such as 'A', 'B', or 'L' means for me either some integral term or another integral proposition.

When a single term is put for several ones, the latter are the definition or the assumed value, and the former is the defined value, for example: if for AB I put C, i.e. when A = BC is a primitive proposition.

A and B coincide if, by substitution of the assumed values in place of the terms and conversely, the same thing, i.e. the same formula, appears on both sides.⟩

I say that statements *coincide* if one can be substituted for the other without loss of truth, i.e. if they infer each other reciprocally.

(1) The (direct) statement 'L' and the (reflexive) statement 'L is *true*' coincide. Hence, 'it is true that L is true (false)' and 'L is true (false)' coincide.[f]

⊙ 'L is true' and 'it is false that L is *false*' coincide.

'That L is false is true' and 'that L is false' coincide. I can demonstrate this as a theorem in this way: 'that L is false' is a statement which may be called M. Now, 'M is true' and 'M' coincide (by 1). Therefore, putting in the place of M its value, 'it is true that L is false' and 'L is false' coincide.

The same is demonstrated in another, though more prolix way, ⊙ also being employed. 'That L is false is true' coincides with this: 'that it is false that L is false is false' (by ⊙), and that again with the sentence 'that it is false that L is true' (also by ⊙), and that, finally, with the sentence that L is false (by 1).

---

[f] It is better to postpone all these remarks until after the propositions have been explained.
    In general, even though A is a term, one can always say 'A is true'; 'it coincides with something true'.

(2) Si coincidunt A et B, coincidunt etiam non-A et non-B.

A non-A contradictorium est.

⟨Possibile est quod non continet contradictorium seu A non-A.

Possibile est quod non est: Y, non-Y.⟩

⟨Dico aliquid impossibile esse seu contradictionem continere, sive terminus sit incomplexus continens A non-A sive sit propositio quae rursus vel dicat coincidere ea quorum unum continet contradictorium alterius, vel contineat terminum incomplexum impossibilem; nam quoties coincidere dicuntur ea quorum unum continet contradictorium alterius utique idem continet terminum contradictorium. Quoties aliquid continet id cujus contradictorium continet, utique continet terminum contradictorium. Itaque adhibita propositione impossibili prodit terminus contradictorius incomplexus.⟩

Coincidunt Non-Non-A et A; adeoque, si coincidunt Non-A et B, coincident etiam non-B et A. ⟨Si A = B, etiam non-A = non-B.

Si A = quoddam verum. Ergo non-A = non $\overline{quoddam\ verum}$ seu nullum verum seu falsum, nam non-A continet non AY.⟩[8]

(3) Coincidunt Non verum et falsum.

Ergo et coincident non-falsum, et Verum.[9]

(4) Coincidunt L esse veram est vera, et L esse non veram est non vera, adeoque coincidunt L[10] et L esse falsam est falsa. Nam L idem est quod L est vera, et haec idem quod L esse veram est vera (per 1), et haec idem quod L non esse veram est non vera (per 4). Et haec idem quod L esse falsam est falsa (per 3).

Coincidunt L, et L esse non falsam est non falsa. Nam L idem est quod L esse veram est vera (per 1) et haec idem est quod L esse non falsam est non falsa (per 3).

Coincidunt L esse falsam et L esse non veram est non falsa.

Coincidunt L esse falsam et L esse non falsam est non vera.

Haec facile demonstrantur ex praecedentibus.

---

[8] ['Ergo' is missing in the Academy edition].

[9] Haec omnia intellige si Termini sint possibiles, nam alioqui neque verum neque falsum in propositionibus quas ingrediuntur locum habet.

[10] [L: coincidunt L; et L esse falsam].

(2) If A and B coincide, then non-A and non-B also coincide.

'A non-A' is a *contradiction*.

⟨*Possible* is that which does not contain a contradiction, i.e. 'A non-A'. Possible is that which is not: 'Y, non-Y'.⟩

⟨I say that something is impossible, i.e. that it contains a contradiction, whether it is an incomplex term containing 'A not-A' or a proposition which either says again that those things coincide of which the one contains the contradictory of the other or contains an impossible incomplex term.* For as often as those things of which one contains a contradictory of the other are said to coincide, surely this assertion contains a contradictory term. As often as something contains that of which it contains the contradictory, surely it contains a contradictory term. Thus, when an impossible proposition is employed, a contradictory incomplex term emerges.⟩

Non-non-A and A coincide; thus, if Non-A and B coincide, non-B and A will also coincide.

⟨If A = B, then non-A = non-B as well.

If A = something true, then non-A = non-*something true,* that is, nothing true, i.e. false, for non-A contains non AY.⟩*

(3) 'Not true' and 'false' coincide.

Therefore, 'non-false' and 'true' will also coincide.[g]

(4) 'It is true that L is true' and 'it is not true that L is not true' coincide; hence L and 'that L is false is false' also coincide. For L is the same as 'L is true', and this is the same as 'it is true that L is true' (by 1), and this is the same as 'it is not true that L is not true' (by 4). And this is the same as 'it is false that L is false' (by 3).

L and 'it is not false that L is false' coincide. For L is the same as 'it is true that L is true' (by 1), and this is the same as 'it is not false that L is not false' (by 3).

'L is false' and 'it is not false that L is not true' coincide.

'L is false' and 'it is not true that L is not false' coincide.

These are easily demonstrated from what precedes.

---

[g] Of course, all this becomes intelligible if terms are possible, because otherwise neither the true nor the false has any place in the propositions into which these terms enter.

Generaliter, si propositio vera aut non vera, falsa aut non falsa dicatur, verum in verum, falsum in falsum facit verum. Non in non aequipollet omissioni utriusque.

Demonstratur etiam ex his omnem propositionem aut veram aut falsam esse. Seu si L est non vera, est falsa; si est vera est non falsa; si est non falsa est vera; si est falsa est non vera. Omnia per 3.

Propositiones autem 1, 2, 3, 4 faciunt officium definitionum, unde sine probatione sunt assumtae, indicant enim usum quorundam signorum nempe veritatis et falsitatis, affirmationis et negationis.

Propositio est A coincidit ipsi B, A est B (seu ipsi A inest B, seu ipsi A substitui potest B); A non coincidit ipsi B.

A autem et B significare possunt Terminos, vel propositiones alias.

(5) A non coincidit ipsi B idem est ac A coincidere ipsi B est falsum.

(6) Si A coincidit ipsi B, B coincidit ipsi A.

(7) Si A non coincidit ipsi B, B non coincidit ipsi A.

(8) Si A coincidit ipsi B, et B coincidit ipsi C, etiam A coincidit ipsi C.

(9) Si A coincidit ipsi B; non-A coincidit ipsi non-B.

Haec quatuor axiomata sunt corollaria hujus definitionis quod coincidunt, quorum unum alteri substitui potest.

(10) *Propositio per se vera* est A coincidit ipsi A.

(11) *Propositio per se falsa* est A coincidit ipsi non-A.

(12) Hinc colligitur falsum esse non-A coincidere ipsi A (per 6).

(13) Item colligitur verum esse A non coincidere ipsi non-A (per 5). Hae propositiones possent referri ad veras per consequentiam.

Porro A ut dixi hoc loco significat Terminum vel propositionem, hinc non-A significat contradictori[u]m termini vel contradictoriam propositionis.

(14) Si propositio ponatur, nec adjicitur aliud, intelligitur esse veram, coincidit cum 1.

(15) Non-B coincidit ipsi non-B est corollarium ipsius 10, posito Non-B coincidere [ipsi] A.

In general, if a given proposition is called true or not true, false or not false, then true applied to true, false applied to false, makes true. 'Non' applied to 'non' is equivalent to the omission of both.

From these it is also demonstrated that every proposition is either true or false, i.e. if L is not true, it is false; if it is true, it is not false; if it is not false, it is true; if it is false, it is not true. All this follows from 3.

Propositions 1, 2, 3, 4 play the role of definitions, therefore, they have been assumed without proof, because they show how to employ certain signs, namely of truth and falsity, affirmation and negation.

A *Proposition* is 'A coincides with B', 'A is B' (i.e. 'B inheres in A', i.e. 'B can be substituted at the place of A'); 'A does not coincide with B'.

A 'A' and B 'B', however, can signify terms, or alternatively other propositions.

(5) 'A does not coincide with B' is the same as 'it is false that A coincides with B'.

(6) If A coincides with B, B coincides with A.

(7) If A does not coincide with B, B does not coincide with A.

(8) If A coincides with B and B coincides with C, then A also coincides with C.

(9) If A coincides with B, non-A coincides with non-B.

These four axioms* are corollaries of the definition according to which those things coincide of which the one can be substituted for the other.

(10) A *proposition true in itself* is 'A coincides with A'.

(11) A *proposition false in itself* is 'A coincides with non-A'.

(12) From this it is inferred that it is false that non-A coincides with A (by 6).

(13) Likewise it is inferred that it is true that A does not coincide with non-A (by 5). These propositions could be traced back to the true ones by means of a consequence.

Moreover, as I said, 'A' here means either a term or a proposition; therefore, 'non-A' means either the contradictory of a term or the contradictory of a proposition.

(14) If a proposition is stated and nothing added, it is understood to be true (this coincides with 1).

(15) Non-B coincides with non-B: it is a corollary of 10, once we assume that non-B coincides with A.

(16) *Propositio Affirmativa* A est B sive A continet B, seu (ut loquitur Aristoteles) ipsi A inest B (in recto scilicet). Hoc est si pro A substituatur valor prodibit: A coincidere ipsi BY. Ut homo est animal, seu homo idem est quod Animal..., nempe Homo idem est quod Animal rationale. Nota enim Y significo aliquod incertum, ut proinde BY idem sit quod quoddam B seu Animal...(ubi subintelligitur rationale, si modo sciamus quid subintelligendum sit) seu quoddam animal. Itaque A est B idem est quod A esse coincidens cuidam B seu A = BY.

⟨Notabile est pro A = BY posse etiam dici A = AB et ita non opus est assumtione novae literae. Praesupponit autem haec notatio quod AA idem est quod A, oritur enim redundantia.⟩

(17) Hinc coincidunt: A esse B, et quoddam B coincidere ipsi A, seu BY = A.

(18) Coincidunt A et AA, et AAA, etc. ex natura hujus characteristicae, seu Homo, et Homo Homo, et Homo homo homo. Itaque si quis dicatur esse Homo pariter et animal, resolvendo Hominem in animal rationale, pariter dicetur Animal rationale et Animal, id est animal rationale.

⟨Hinc patet etiam ex AC = ABD non licere inferri C = BD, patet enim et in A = AB non posse utrinque omitti A. Si ob AC = ABD inferri posset C = BD, praesupponendum esset, nihil quod continetur in A contineri et in C quin contineatur et in BD, et contra.⟩

(19) Si A sit B, pro A poni potest B, ubi tantum de continendo agitur, ut si A sit B et B sit C, A erit C. Demonstratur ex natura coincidentiae, nam coincidentia substitui sibi possunt (nisi in propositionibus quas dicere possis formales, ubi unum ex coincidentibus ita formaliter assumitur, ut ab aliis distinguatur, quae revera sunt reflexivae, et non tam de re loquuntur, quam de nostro concipiendi modo, ubi utique discrimen est).[11] Itaque cum (per 16) A = BY et B = CZ, Ergo A = CYZ seu A continet C.[12]

---

[11] NB.

[12] Licebit et habere generale quoddam indefinitum, quasi Ens quoddam seu quoddam, ut in communi sermone, tunc nulla orietur coincidentia.

(16) An *Affirmative proposition* is 'A is B' or 'A contains B' or, as Aristotle says, 'B is in A' (i.e. in the nominative case).* That is, if we substitute a value for A, we will have: 'A coincides with BY'. For example, 'man is an animal', i.e. 'man is the same as animal...', namely 'man is the same as a rational animal'. With the sign 'Y', indeed, I mean something undetermined, so that 'BY' is the same as 'some B' or '...animal' (where 'rational' is implied, provided that we know what is to be implied) or 'some animal'. Thus, 'A is B' is the same as 'A is coincident with some B' or A = BY. ⟨It is worth noting that instead of A = BY it is also possible to say A = AB, and so there is no need for the assumption of a new letter. This notation, however, presupposes that AA is the same as A, for a redundancy arises⟩.

(17) Hence these coincide: 'A is B' and 'some B coincides with A', i.e. BY = A.

(18) Due to the nature of this characteristic A, AA, and AAA, etc., i.e. 'man', and 'man man', and 'man man man' coincide. Therefore, if someone is said to be both a man and an animal, by analyzing 'man' into 'rational animal', he will be called at the same time 'rational animal' and 'animal', i.e. 'rational animal'.

⟨From this is also clear that from AC = ABD we are not allowed to infer C = BD; for it is clear that in A = AB we cannot omit A on both sides. If from AC = ABD one could infer that C = BD, it would have to be presupposed that nothing contained in A is also contained in C unless it is also contained in BD, and conversely.⟩

(19) If A is B, B can be put in place of A, where only containing is involved. If, for instance, A is B and B is C, A will be C. This is demonstrated by the nature of coincidence: things that coincide, indeed, can be substituted for one another (except in the case of propositions which you could call 'formal', where one of the coinciding things is taken formally in such a way that it is distinguished from the others; these propositions, however, are reflexive, and are asserted not so much about a thing as about our way of conceiving it, where there is certainly a distinction between them).[h]* Thus, given that (by 16) A = BY and B = CZ, therefore A = CYZ i.e. A contains C.[i]

[h] NB.
[i] We may also have some general indefinite term, such as 'some being' or 'something', as in ordinary speech, and then no coincidence will arise.

(20) Notandum est quod in hoc calculo fuerat praemittendum; pro quotlibet literis simul poni posse unam, ut YZ = X, sed nondum usurpatam in hoc calculo Rationis, ne oriatur confusio.

(21) Deinde definitas a me significari prioribus Alphabeti literis, indefinitas posterioribus, nisi aliud significetur.

(22) Pro quotcunque definitis substitui posse unam definitam, cujus valor seu definitio sunt illae pro quibus substituta est.

(23) Pro qualibet definita substitui posse indefinitam nondum usurpatam. Ac proinde et pro quotlibet definitis, et pro definitis et indefinitis, seu poni potest A = Y.

(24) Cuilibet literae adjici potest nova indefinita, ut pro A poni potest AY, nam A = AA (per 18) et A est Y (seu pro A poni potest Y, per 23). Ergo A = AY.

(25) A esse B (A continere B) infert (continet) quoddam B esse (continere) A.

Nam $\overline{A\ esse\ B} = \overline{BY = A}$ (per 17) = $\overline{BY = AY}$ (per 24) = quoddam B esse A (per 17).

(26) Admonenda adhuc quaedam circa hunc calculum, quae praemittere debueramus. Nempe quod de quibuslibet literis nondum usurpatis asseritur generaliter vel concluditur, non tanquam Hypothesis, id de quotlibet aliis literis intelligi. Itaque si asseratur A = AA, etiam dici poterit B = BB.

(27) Quoddam B = YB. Itaque similiter qu. A = ZA. Nimirum licet hoc quidem dicere ad imitationem prioris (per 26) sed nova assumenda est indefinita pro posteriori aequatione, nempe Z, ut paulo ante fuerat Y.

(28) Terminus simpliciter positus a me solet usurpari pro universali, ut A est B, id est omne A est B, seu in notione A continetur notio B.

(29) A est B, Ergo quoddam A est B (sive A continere B, infert seu continet quoddam A continere B). Nam $\overline{A\ est\ B}$ = $\overline{AY\ est\ B}$ (per 24).

(20) Something should be noted here that should have been stated earlier in this calculus: that one letter can be put in place of any number of other letters together; for example: YZ = X. But this has not yet been employed in this calculus of reason, lest confusion arise.*

(21) Moreover, I use the earlier letters of the alphabet to denote definite letters and the later letters to denote the indefinite ones, unless otherwise indicated.

(22) For any number whatsoever of definite letters one can substitute one definite letter, whose value or definition is those letters for which it has been substituted.

(23) An indefinite letter not yet used can be substituted for any definite letter. Similarly, one can perform the substitution for any number of definite letters, and for definite and indefinite ones as well, i.e. one can put A = Y.

(24) To any letter a new indefinite one can be added; thus, for A we can put AY, because AA = A (by 18) and A is Y (or for A we may put Y, by 23). Therefore A = AY.

(25) That A is B (that A contains B) entails (contains) that some B are (contain) A.

For $\overline{\text{A is B}}$ = $\overline{\text{BY} = \text{A}}$ (by 17) = $\overline{\text{BY} = \text{AY}}$ (by 24) = some B are A (by 17).*

(26) Here we should mention some things concerning this calculus that we should have stated earlier, namely that what is asserted in general or concluded, not as hypothesis, about any letters whatsoever not yet used is to be understood of any number of other letters. Thus, if A = AA, it will also be possible to say B = BB.

(27) Some B = YB. Thus, similarly some A = ZA. Surely this too may be said in imitation of the former (by 26), but we need to assume a new indefinite letter, namely Z, for the latter equation, just as Y had been assumed a little earlier.

(28) A term set forth by itself* I customarily use for a universal as, for instance, 'A is B', i.e. 'every A is B', that is, the notion of B is contained in the notion of A.

(29) 'A is B', therefore 'some A is B' (i.e. that $\underline{\text{A contains B}}$ entails or contains that some A contains B). For $\overline{\text{A is B}}$ = $\overline{\text{AY is B}}$ (by 24).*

(30) A esse B et B esse A idem est quod A et B coincidere, sive A coincidere ipsi B quod coincidit ipsi A. Nam A = BY et B = AZ. Ergo (per [19]¹³) A = AYZ. Ergo Y, Z sunt superfluae seu Z continetur in A. Ergo pro B = AZ dici potest B = A.

(31) Scilicet notandum et hoc est, si A = AY, tunc vel Y est superfluum, vel potius generale ut Ens, et utique impune omitti potest, ut Unitas in multiplicatione apud Arithmeticos, vel Y inest in A. Imo revera semper inest Y in A, si dicatur A = YA.¹⁴

(32) *Propositio Negativa.* A non continet B, seu A esse (continere) B falsum est seu A non coincidit BY.

⟨Si B sit propositio, non-B idem est quod B est falsum seu τὸ B esse falsum. Non-B, intelligendo B de propositione in materia necessaria, vel est necessarium vel impossibile. At secus est in incomplexis.

Notionem sumo tam pro incomplexa quam complexa. Terminum pro incomplexa categorematica.⟩

(32[⟨bis⟩]) B, non-B est impossibile, seu si B non-B = C, erit C impossibile.

⟨Impossibile in incomplexis est non-Ens, in complexis est falsum⟩.

(33) Hinc si A = non-B, erit AB impossibile.

(34) Quod continet B, non-B idem est quod *impossibile* seu EB, non-B idem est quod impossibile.

(35) *Propositio falsa* est, quae continet AB continere non-B (posito B et A esse possibiles). Intelligo autem B et [A]¹⁵ tam de Terminis, quam de Propositionibus.

⟨A continere B et A continere C idem est quod A continere BC. Hinc si A continet B, etiam continet AB. Hinc si AB continet non-B, etiam AB continebit AB non-B.⟩

(36) A = B. Ergo A est B, seu A = B, continet quod A est B. Nam si Y sit superflua, fiet A = BY, id est A = B. Idem aliter demonstratur: A = B idem est quod A = BY et B = AY. Ergo A = B continet A = BY. Item A = B, ergo AA = BA. Ergo A = BA. Ergo A est B.

---

¹³ [L: 31].   ¹⁴ NB.   ¹⁵ [L: Y].

(30) That A is B and B is A is the same as that A and B coincide, i.e. that A coincides with B, which coincides with A. For A = BY and B = AZ. Therefore, (by [19]) A = AYZ. Therefore, Y, Z are superfluous, i.e. Z is contained in A. Therefore, in place of B = AZ we may say B = A.

(31) Indeed, we must also observe this: if A = AY, then either Y is superfluous or rather something general, like being, and in both cases it can be safely omitted, like unity in the case of arithmetical multiplication, or Y inheres in A; better, Y is always in A if we say A = YA.[j]

(32) A *negative proposition*: 'A does not contain B', i.e. 'it is false that A is (contains) B', i.e. 'A does not coincide with BY'.

⟨If B is a proposition, then 'non-B is the same as B' is false, i.e. B's being false. Non-B, understanding 'B' as referring to a proposition in the domain of the necessary, is either necessary or impossible. Things, however, are different in the case of incomplex terms.

I assume a notion to be incomplex or complex, and a term an incomplex categorematic notion.⟩

(32[⟨bis⟩]) 'B non-B' is impossible, i.e. if B non-B = C, C will be impossible.

⟨In incomplex notions what is impossible is a non-being; in the complex notions it is false.⟩

(33) Hence, if A = non-B, AB will be impossible.

(34) That which contains B non-B is the same as *impossible*, i.e. EB non-B is the same as 'impossible'.

(35) A *false proposition* is one which contains that AB contains non-B (assuming that A and B are possible). I understand B and A as playing the role of terms and of propositions as well.

⟨That A contains B and A contains C is the same as that A contains BC. Hence, if A contains B, it also contains AB. Hence, if AB contains non-B, AB too will contain AB non-B.⟩

(36) A = B. Therefore, A is B, i.e. A = B contains that A is B. For if Y is superfluous, we shall have A = BY, i.e. A = B. The same is demonstrated in another way: A = B is the same as A = BY and B = AY. Therefore, A = B contains A = BY. Likewise, A = B, therefore AA = BA. Therefore, A = BA. Therefore, A is B.

---

[j] NB.

(37) B est B, nam B = B (per 10). Ergo B est B (per 36).

(38) AB est B. Est indemonstrabilis, et sive identica sive definitio est, vel τοῦ Est, vel continentis, vel ⟨verae propositionis⟩. Nam significatur AB, seu id quod continet B, esse B seu continere B.

(39) Si B continet C, tunc AB continet C. Nam AB est B (per 38) B est C (ex hypothesi). Ergo (per 19) AB est C.

(40) *Vera propositio* est quae coincidit cum hac: AB est B, seu quae ad hanc primo veram reduci potest. (Puto id et ad non-categoricas applicari posse.)

(41) Igitur cum falsa sit quae non est vera (per 3) sequitur (ex 40) falsam propositionem idem esse quod propositionem quae non coincidit cum hac: AB est B, seu falsam propositionem idem esse quod propositionem quae non potest probari. Propositiones facti non semper probari possunt a nobis, et ideo assumuntur ut Hypotheses.

(42) A continet B et A non continet B, earum una est vera altera falsa seu sunt *Oppositae*, nam si una probari potest altera non potest, ⟨modo termini sint possibiles. Ergo (per 41) non simul verae sunt aut falsae.⟩

(43) B continere non-B est falsa seu [B] non continet non-B, patet ex praecedenti. Nam utcunque resolvas manet semper haec forma, nunquam fiet AB est B. Patet et aliter. B continet B (per 37). Ergo non continet non-B alioqui foret impossibilis (per 32).

⟨Falsum esse B continere non-B, intelligendum est et de propositione B, quae non continet contradictoriam.⟩

(44) Non-B continere B est falsa, patet eodem modo.

(45) B et non-B coincidere est falsa. Patet ex 43 et 44. Supponunt autem haec terminum B esse possibilem.

(46) AB continere non-B est falsa, seu AB non continet non-B. Suppono autem AB esse possibilem. Demonstratur ut 43. Nam AB continet B, ergo non continet non-B, quia est non impossibilis (per 32).[16]

---

[16] Cavendum est ne syllogismis utamur, quos legitimos esse nondum demonstravimus.

(37) B is B, for B = B (by 10). Therefore, B is B (by 36).

(38) AB is B. This cannot be demonstrated and is either an identical proposition or a definition, either of 'is', or of 'contains', or ⟨of a true proposition⟩. For it means that AB, i.e. that which contains B, is B, i.e. contains B.

(39) If B contains C, then AB contains C. For AB is B (by 38), B is C (*ex hypothesi*), therefore AB is C (by 19).

(40) A *true proposition* is one that coincides with 'AB is B' or which can be reduced to this primary truth (I think that this can be applied to non-categorical propositions as well).

(41) Therefore, because a false proposition is one which is not true (by 3), it follows (from 40) that a false proposition is the same as a proposition that does not coincide with 'AB is B', i.e. a false proposition is the same as a proposition which cannot be proved. Propositions of fact cannot always be proved by us, and so they are assumed as hypotheses.

(42) 'A contains B' and 'A does not contain B': of these propositions, one is true and the other false, i.e. they are *opposites*. For if one can be proved, the other cannot be, ⟨provided that the terms are possible. Therefore, (by 41) they are not true or false at the same time⟩.

(43) It is clear from what precedes that it is false that B contains non-B, i.e. that B does not contain non-B. For however you analyse them, the same form always remains, and will never become 'AB is B'. This is also clear in another way. 'B contains B' (by 37); therefore, it does not contain non-B; otherwise it would be impossible (by 32).

⟨That it is false that B contains non-B must be understood as referring to the proposition B, which does not contain a contradiction⟩.

(44) In the same way it is clear that it is false that non-B contains B.

(45) It is false that B and non-B coincide. This is clear from 43 and 44. These assume, however, that the term B is possible.

(46) It is false that AB contains non-B, i.e. AB does not contain non-B. I suppose AB to be possible. This is demonstrated in the same way as 43. For AB contains B; therefore, it does not contain non-B, because it is not impossible (by 32).[k]

---

[k] We must take care not to use syllogisms which we have not yet demonstrated to be legitimate.

(47) A continet B est *Universalis affirmativa respectu ipsius A* subjecti.

(48) AY continet B est *Particularis Affirmativa respectu ipsius A.*

(49) Si AB est C, sequitur quod AY est C, seu sequitur quoddam A est C, nam assumi potest B = Y per 23.

(50) AY non est B est *Universalis negativa.*

(51) Hinc sequitur Universalem negativam et *Particularem Affirmativam* esse oppositas, seu si una est vera altera est falsa (ex 48 et 50).

(52) Particularis affirmativa verti potest simpliciter seu si quoddam A est B sequitur quod quoddam B est A. Hoc ita demonstro: AY est B ex Hypothesi, id est (per 16) AY coincidit ipsi BY. Ergo (per 6) BY coincidit ipsi AY. Ergo (per 16) BY est A. Quod erat dem.[17]

(53) Universalis Negativa convertitur simpliciter, seu si Nullum A est B sequitur quod Nullum B est A. Nam AY non est B (ex hypothesi), ergo AY non coincidit BY (per 16). Ergo BY non coincidit AY (per 6). Ergo (per 16) BY non est A. Quod erat dem.

(54) Universalis affirmativa convertitur per accidens, seu si omne A est B sequitur quod quoddam B est A. Nam A est B ex hypothesi. Ergo quoddam A est B (per 29). Ergo (per [52][18]) quoddam B est A. Idem brevius: A coincidit BY (per 16). Ergo BY coincidit A (per 6). Ergo (per 36) BY est A.[19] Operae pretium erit conferre has duas demonstrationes, ut appareat utrum eodem recidant, an vero detegant veritatem alicujus propositionis hactenus sine demonstratione assumtae.[20]

(55) Si A continet B et A est vera, etiam B est vera. Per falsam literam intelligo vel terminum falsum (seu impossibilem, seu qui est non-Ens), vel propositionem falsam. Et per ve[ram] eodem modo intelligi possit terminus possibilis, vel propositio vera. Et ut postea explicatur, totus syllogismus mihi etiam propositio est. Caeterum quod hic assero etiam sic enuntiari potest, quaelibet pars veri est vera seu quod continetur in vero, est verum.[21] Demonstrari potest ex sequenti.

---

[17] Majusculis notentur propositiones fundamentales seu indemonstratae ut L1 (vel simul numeris communibus et diversis).

[18] [L: 53].

[19] Dicendum de collatione horum Nullum A est B et Omne A est non-B. Item de conversione per contrapositionem ipsius Universalis affirmativae. Pro Nullum A est B licebitne dicere Omne A non est B?

[20] NB.   [21] NB.

(47) 'A contains B' is a *universal affirmative with regard to* the subject A.

(48) 'AY contains B' is a *particular affirmative with regard to* A.

(49) If 'AB is C', it follows that 'AY is C', i.e. it follows that 'some A is C', for we may assume, by 23, B = Y.

(50) 'AY is not B' is a *universal negative*.

(51) From this follows that a universal negative and a *particular affirmative proposition* are opposites, i.e. if the one is true, the other is false (by 48 and 50).

(52) A particular affirmative can be converted *simpliciter*, i.e. if some A is B, it follows that some B is A. I demonstrate this as follows: AY is B, *ex hypothesi*, i.e. (by 16) AY coincides with BY. Therefore, (by 6) BY coincides with AY. Therefore, (by 16) BY is A. Q.E.D.[1]

(53) A universal negative is converted *simpliciter*, i.e. if no A is B, it follows that no B is A. For AY is not B (*ex hypothesi*); therefore, AY does not coincide with BY (by 16). Therefore, BY does not coincide with AY (by 6). Therefore, (by 16) BY is not A. Q.E.D.

(54) A universal affirmative is converted by accident, i.e. if every A is B, it follows that some B is A. For A is B *ex hypothesi*. Therefore, some A is B (by 29). Therefore, (by [52]) some B is A. More briefly: A coincides with BY (by 16). Therefore, BY coincides with A (by 6). Therefore, (by 36) BY is A.[m] It will be worth our while to compare these two demonstrations, so that we may see whether they come to the same, or whether they bring into the open the truth of some proposition that had been assumed hitherto without demonstration.[n]

(55) If A contains B and A is true, then B is also true.* By a 'false letter' I understand either a false term (i.e. a term which is impossible or a non-being) or a false proposition. Similarly, a true letter can be understood either as a possible term or a true proposition.[o] And, as is explained later, I consider the whole syllogism a proposition also.* What I am asserting here, however, can be stated this way: any part of the true is true, i.e. what is contained in the true is true.

This can be demonstrated from what follows.

---

[1] Fundamental, i.e. undemonstrated, propositions should be denoted by capital letters, such as 'LI' (or, at the same time, by common and different numbers).

[m] We must say something concerning the comparison between 'No A is B' and 'Every A is non-B'; also concerning the conversion by contraposition of the universal affirmative. Will it be correct to say 'Every A is not B' for 'No A is B'?

[n] NB.   [o] NB.

(56) *Verum* in genere sic definio, *Verum* est A, si pro A ponendo valorem, et quodlibet quod ingreditur valorem ipsius A rursus ita tractando ut A, si quidem id fieri potest, nunquam occurrat B et non-B ⟨seu [contradictio]⟩.[22] Hinc sequitur ut certi simus veritatis vel continuandam esse resolutionem usque ad primo vera ⟨aut saltem jam tali processu tractata, aut quae constat esse vera⟩, vel demonstrandum esse ex ipsa progressione resolutionis, seu ex relatione quadam generali inter resolutiones praecedentes et sequentem, nunquam tale quid occursurum, utcunque resolutio continuetur. Hoc valde memorabile est, ita enim saepe a longa continuatione liberari possumus. Et fieri potest, ut resolutio ipsa literarum aliquid circa resolutiones sequentium contineat, ut hic resolutio Veri. Dubitari etiam potest an omnem resolutionem finiri necesse sit in primo vera seu irresolubilia in primis in propositionibus contingentibus, ut scilicet ad identicas reduci [⟨rursus possint⟩].

(57) *Falsum in genere* definio quod non est verum. Itaque ut constet aliquid esse falsum, vel necesse est ut sit oppositum veri, vel ut contineat oppositum veri, vel ut contineat contradictionem seu B et non-B, vel si demonstretur, utcunque continuata resolutione non posse demonstrari quod sit verum.

(58) Itaque quod continet falsum est falsum.

(59) Potest tamen aliquid continere verum, et tamen esse falsum. Si scilicet (per 58) praeterea falsum contineat.

(60) Videmur etiam hinc discere posse discrimen veritatum necessariarum ab aliis, ut scilicet verae necessariae sint quae ad identicas reduci possunt, aut quarum oppositae reduci possint ad contradictorias; et falsae impossibiles, quae ad contradictorias reduci possint, aut quarum oppositae reduci possint ad identicas.

---

[22] [L: contradictionem].

(56) I define *true* in general this way: A is *true* if, substituting a value in place of A, and treating in the same way as A (if possible) everything that enters into the value of A, there never arises B and not-B, ⟨i.e. a contradiction⟩. From this it follows that to be certain of a truth, either we must continue the resolution until we reach the first truths (⟨either, at least, those which have already been treated by this procedure or those which are clearly recognized as true⟩) or we must demonstrate that from the very progression of the resolution, i.e. from a certain general relation between the preceding resolutions and the one that follows, nothing of this kind will ever appear, however long the resolution is continued. This is particularly worth noting, for in this way we can often be freed from a long continuation of the resolution.* It can also happen that the resolution of letters itself contains something concerning the resolutions of letters which follow, such as the resolution of 'true' here. We may also doubt whether every resolution must necessarily end with primary truths, i.e. with what cannot be further resolved, especially in the case of contingent propositions, where we may doubt that they can, in their turn, be reduced to identical propositions.

(57) *False in general* I define as that which is not true. And so in order to ascertain that something is false, either it is necessary that it should be the opposite of a truth, or that it should contain the opposite of a truth, or that it should contain a contradiction, i.e. B and non-B, or if it should be demonstrated that, however long the resolution is continued, it cannot be demonstrated that it is true.*

(58) Therefore, what contains the false is false.

(59) Something, however, can contain a truth and yet be false, namely, if (by 58) it also contains something false.

(60) It also seems that from this we can learn the distinction between necessary truths and others, namely that necessary truths are those that can be reduced to identical propositions, or whose opposites can be reduced to contradictory propositions, and that impossible propositions are false which can be reduced to contradictory propositions, or whose opposites can be reduced to identical propositions.

(61) Possibiles sunt de quibus demonstrari potest nunquam in resolutione occursuram contradictionem. Verae contingentes sunt quae continuata in infinitum resolutione indigent. Falsae autem contingentes quarum falsitas non aliter demonstrari potest, quam quod demonstrari nequeat esse veras. Videtur esse dubium, utrum sufficiat ad demonstrandam veritatem, quod continuata resolutione certum sit nullam occursuram esse contradictio[nem][23]. Inde enim sequetur omne possibile esse verum. Equidem Terminum incomplexum qui est possibilis, voco verum, et qui est impossibilis, voco falsum. At de Termino complexo, ut: A continere B, seu A esse B, ambigi potest. Resolutionem autem termini complexi intelligo in alios terminos complexos. Scilicet sit $\overline{\text{A esse B}}$ = L et sit B = CD et $\overline{\text{A esse C}}$ = M, et $\overline{\text{A esse D}}$ = N, utique fiet: L = MN. Licet autem subjectum A resolvatur, non potest pro A substitui pars valoris, sed substituendus est valor integer, quod obiter moneo. Et si C = EG et D = FG, et A = EFG, poterit M resolvi in has duas $\overline{\text{A = EFG}}$ = P et $\overline{\text{EFG = EG}}$ = Q, seu erit M = PQ; et similiter N in has duas resolvi poterit: $\overline{\text{A = EFG}}$ = P, et $\overline{\text{EFG = FG}}$ = R, ergo L = PQR, quae sunt primo verae, nam P est Hypothesis, Definitio scilicet vel experimentum, R et Q sunt axiomata prima. Verum si porro pergamus, requiritur ad definitionem, ut constet eam esse possibilem, seu necesse est ut demonstretur A esse possibilem, seu ut demonstretur, EFG non involvere contradictionem, id est non involvi X non-X. Quod cognosci non potest nisi experimento, si constet A existere, vel extitisse, adeoque esse possibile aut saltem extitisse aliquid ipsi A simile[24] (Quanquam revera hic casus fortasse non possit dari, nam duo completa nunquam sunt similia, et de incompletis sufficit unum ex duobus similibus existere, ut incompletum, id est denominatio communis, possibilis dicatur (imo tamen videtur esse utile, seu si sphaera una extitit, dici poterit recte quamlibet sphaeram esse possibilem)).[25]

---

[23] [L: contradictio].
[24] Cujus simile possibile est, id ipsum est possibile.   [25] NB.

(61)[p] Possible propositions are those of which it can be demonstrated that a contradiction will never arise in their resolution. Contingent truths are those, which need a resolution continued to infinity. False contingent propositions, on the other hand, are those whose falsity cannot be demonstrated otherwise than by proving that it is impossible to demonstrate that they are true. It seems doubtful whether it is sufficient for proving a truth that, on continued resolution, it should be certain that no contradiction would arise. From this it follows, indeed, that everything which is possible is true. For my part, I call 'true' an incomplex term which is possible, and 'false' an impossible one. But we may have some doubts concerning a complex term like 'A contains B', or 'A is B'. I conceive of the resolution of a complex term as a resolution into other complex terms. Given, for instance, $\overline{\text{A is B}} = \text{L}$, and $\text{B} = \text{CD}$ and $\overline{\text{A is C}} = \text{M}$ and $\overline{\text{A is D}} = \text{N}$, we shall have $\text{L} = \text{MN}$. If, however, the subject A is resolved, part of its value cannot be substituted for A, but its whole value must be substituted, as I remind the reader in passing. And if $\text{C} = \text{EG}$ and $\text{D} = \text{FG}$, and $\text{A} = \text{EFG}$, then M can be analysed into these two: $\overline{\text{A} = \text{EFG}} = \text{P}$ and $\overline{\text{EFG} = \text{EG}} = \text{Q}$, i.e. $\text{M} = \text{PQ}$: and analogously N can be analysed into these two: $\overline{\text{A} = \text{EFG}} = \text{P}$ and $\overline{\text{EFG} = \text{FG}} = \text{R}$; therefore, $\text{L} = \text{PQR}$, which are first truths, because P is a hypothesis, a definition namely or something given by experience, while R and Q are first axioms. If we proceed further, however, a requisite for a definition is that it should be established that it is possible; that is, it must be demonstrated that A is possible, i.e. that EFG does not involve a contradiction, i.e. that X non-X is not involved. And this can be known only by experience, if it is established that A exists, or has existed, and therefore that it is possible or, at least, that something similar to A has existed.[q] (This case, however, perhaps cannot really arise, since two complete things are never similar, and regarding incomplete things it is sufficient that only one of two similars exists in order for the incomplete thing, i.e. the common denominator, to be called possible (this, however, seems to be useful, i.e. if a single sphere had existed, then it could rightly be said that any sphere is possible)).[r]

---

[p] [The beginning of this paragraph has been heavily corrected and on the margin of the manuscript Leibniz has first written and then struck out the words: 'These things are bad, but have been corrected later'].

[q] If something similar to something else is possible, then the latter is also possible.

[r] NB.

Unde patet rem eodem modo procedere in Terminis complexis et in incomplexis. Nam probare verum esse terminum complexum est eum reducere in alios terminos complexos veros et hos tandem in terminos complexos primo veros, hoc est, in axiomata (seu propositiones per se notas), definitiones terminorum incomplexorum quos probatum est esse veros; et experimenta. Similiter Terminos incomplexos esse veros probatur reducendo eos in alios terminos incomplexos veros, et hos tandem in alios terminos incomplexos primo veros, hoc est in terminos per se conceptos, vel in terminos, aut terminos quos sumus experti (aut quorum similes sumus experti, ⟨quanquam id adjici opus non sit, nam demonstrari potest uno similium existente possibili et alia esse [possibilia]²⁶⟩). Ita ut omnis resolutio tam complexorum quam incomplexorum, desinat in axiomata, terminos per se conceptos, et experimenta. Fit autem haec resolutio pro quolibet substituendo valorem, nam et cum pro continente substituitur contentum valor substituitur indefinitus, ut sup. n. 16 ostendimus.

(62) Omnis autem propositio vera potest probari. Unde cum experimenta rursus sint propositiones verae, ideo si nullus alius datur probandi modus quam paulo ante descriptus, sequitur rursus experimenta resolvi posse in axiomata, terminos per se conceptos et experimenta, nulla autem dari possunt Experimenta prima, nisi sint ipsa per se nota, seu axiomata.

(63) Quaeritur an experimenta resolvi possint in alia experimenta in infinitum, et omissa mentione experimentorum an possibile sit ⟨quandam probationem esse talem, ut comperiatur⟩ propositionis probationem, semper praesupponere probationem alterius propositionis, quae non sit axioma nec definitio, adeoque rursus indigeat probatione. Unde et necesse est terminos quosdam incomplexos continue ita resolvi posse, ut nunquam deveniatur ad per se conceptos. Alioqui resolutione absoluta apparebit utrum coincidentia virtualis fiat formalis seu expressa sive an res redeat ad identicam.

---

²⁶ [L: similia]. [Couturat (*Opuscules*: 373) suggests changing it to 'possibilia', which, indeed, is quite reasonable: if we leave 'similia', the sentence becomes a platitude (things belonging to a set of similar things are similar). Schupp (1982: 52) proposes 'uno similium existente, possibili[a] et alia esse similia', but in the manuscript it is clear that there is no comma after 'existente' and, moreover, the neuter plural of 'possibilis' makes the syntax of the sentence quite odd.]

From this it is evident that we have the same procedure in the case of complex terms as in that of the incomplex ones. For to prove that a complex term is true is to analyse it into other complex terms, and these finally into other complex terms which are first truths, that is, into axioms (i.e. propositions which are known through themselves), definitions of incomplex terms that have been proved to be true, and data of experience. Analogously, incomplex terms are proved to be true by analysing them into other true incomplex terms, and these finally into other incomplex terms which are first true terms, i.e. either into terms conceived by themselves, or into terms or else into terms which we have experienced (or into terms similar to these, which we have experienced, ⟨although there is no need to add this yet, because it can be demonstrated that if only one out of a number of similar things is possible, then all the other similar things are [possible] as well⟩). Thus every resolution of both complex and incomplex terms ends in axioms, terms conceived by themselves, and data of experience. This resolution is made by substituting a value for any term, because even when the contained is substituted for the containing, what is substituted is an indefinite value, as we have shown above, n. 16.

(62) Every true proposition, however, can be proved. Therefore, since the data of experience are again true propositions, if there is no other means of proof than that described just previously, it follows again that the data of experience can be resolved into axioms, terms conceived by themselves and data of experience. Yet there cannot be primary data of experience unless they are known through themselves, i.e. are axioms.

(63) We are inquiring whether the data of experience can be resolved into others *ad infinitum*; and whether it is possible, without mentioning any data of experience, ⟨for some proof to be such that we find out⟩ that the proof of the proposition always presupposes the proof of another, which is neither an axiom nor a definition, thus being in need of proof again. Therefore, it is also necessary that some incomplex terms can be continually resolved in such a way that we never reach terms which are conceived through themselves. Otherwise, once the resolution has been completed, it will appear whether the virtual coincidence becomes formal or explicit, i.e. whether a reduction has been made to an identical proposition.

(64) Quaeritur igitur an possibile sit resolutionem terminorum incomplexorum aliquando posse continuari in infinitum, ut nunquam perveniatur ad per se conceptos. Et sane si nullae darentur in nobis notiones per se conceptae, quae distincte attingi possint, aut non nisi una ⟨(v.g. notio Entis)⟩; sequitur nec propositionem ullam ratione perfecte demonstrari posse; nam licet ex positis definitionibus et axiomatibus perfecte possit demonstrari sine experimentis, definitiones tamen praesupponunt terminorum possibilitatem, adeoque vel resolutionem in per se conceptos, vel in experimento compertos, reditur ergo ad experimenta seu ad alias propositiones.

(65) Quodsi dicamus possibilem esse continuationem resolutionis in infinitum, tunc saltem observari potest, progressus in resolvendo an ad aliquam regulam reduci possit, unde et in terminorum complexorum, quos incomplexi in infinitum resolubiles ingrediuntur, probatione talis prodibit regula progressionis.

(66) Quodsi jam continuata resolutione praedicati et continuata resolutione subjecti, nunquam quidem demonstrari possit coincidentia, sed ex continuata resolutione et inde nata progressione ejusque regula saltem appareat nunquam orituram contradictionem, propositio est possibilis. Quodsi appareat ex regula progressionis in resolvendo eo rem reduci, ut differentia inter ea quae coincidere debent, sit minor qualibet data, demonstratum erit propositionem esse veram,[27] sin apparet ex progressione tale quid nunquam oriturum, demonstratum est esse falsam, scilicet in [⟨contingentibus⟩].[28] [29]

---

[27] NB.

[28] [L: necessariis]. [In the manuscript Leibniz has clearly written 'in necessariis' (i.e. 'in the case of necessary propositions'), and this version appears in *Opuscules*: 374, Schupp (1989: 56), and Rauzy (1998: 242). The Academy's correction seems to be appropriate, if we consider what Leibniz writes further at §§131–5.].

[29] Dubium: utrum verum omne quod non potest probari falsum; an falsum omne quod non potest probari verum; quid ergo de illis, de quibus neutrum? Dicendum est semper probari posse et verum et falsum resolutione in infinitum saltem. Sed tunc est contingens, seu possibile est ut vera sit, aut ut falsa; idemque est de notionibus, ut in resolutione in infinitum appareant verae aut falsae, id est ad existendum admittendae, vel non. Hoc modo an notio vera erit existens; falsa non existens. Omnis notio impossibilis est falsa, sed non possibilis est vera, itaque falsa erit quae nec est nec erit, ut falsa est talis propositio; etc. Nisi forte malimus nullam existentiae in his habere rationem, et notio vera hic idem quod possibilis; falsa idem quod impossibilis, nisi quando dicitur, v.g. *Pegasus existens*.

(64) The question is, therefore, whether it is possible that the resolution of incomplex terms can sometimes be continued *ad infinitum*, so that we never arrive at terms conceived through themselves. Certainly, if there are in us no notions conceived through themselves which can be grasped in a distinct way, or only one ⟨(e.g. the notion of being)⟩, then it follows that no proposition can be demonstrated perfectly by reason. For although it can be demonstrated perfectly without the data of experience, from the definitions and axioms that have been assumed, nevertheless the definitions presuppose the possibility of the terms, and thus, in turn, they presuppose their resolution either into terms conceived through themselves or into those concepts that have been discovered by means of experience; thus, we come back to the data of experience or to other propositions.

(65) But if we say that the continuation of the resolution *ad infinitum* is possible, then it can at least be noticed whether our progress in resolution can be reduced to some rule; hence such a rule of progression will appear in the proof of complex terms, which have as ingredients incomplex terms susceptible of being analysed *ad infinitum*.

(66) But if, once the resolution of the predicate and of the subject has been continued, a coincidence can never be demonstrated; yet it does at least appear from the continued resolution, and from the progression and its rule arising from it, that a contradiction never emerges, then the proposition is possible. But if, on resolving it, it appears from the rule of progression that we have reached a point where the difference between what should coincide is less than any given difference, then it will have been demonstrated that the proposition is true; otherwise, if it appears from the progression that nothing of this sort will ever arise, then it has been demonstrated that the proposition is false, that is to say, in the case of [contingent] propositions.[s]

---

[*] A doubtful point: is everything true which cannot be proved false, or everything false which cannot be proved true? Then what about those cases where neither can be proved? It must be said that it is always possible to prove both truth and falsity, at least, by a resolution carried *ad infinitum*. But then it is contingent, i.e. it is possible that it is true or that it is false; and the same holds for notions, namely that by means of a resolution carried *ad infinitum* they are shown to be true or false, i.e. to be admitted to existence or not. In this way, will a true notion be existent and a false one non-existent? Every impossible notion is false, but not every possible notion is true, and so that notion will be false which neither exists nor will exist, as a proposition of this kind is false, etc. Unless, perhaps, we prefer to take no account of existence in these cases, and a true notion here will be the same as a possible one, a false notion the same as an impossible one, except when we say, for instance, 'Pegasus existing'.

(67) Necessaria autem propositio est, cujus opposita non est possibilis, seu cujus oppositam assumendo per resolutionem devenitur in contradictionem. Itaque necessaria est quae per identicas demonstrari potest, et definitiones, nullo alio usu experimentorum accedente, quam ut inde constet terminum esse possibilem.

(68) Sed illud adhuc examinandum est, unde sciam me recte progredi in definiendo, nam si dico A = EFG, non tantum scire debeo, E, F, G singula esse possibilia, sed etiam inter se compatibilia, id autem patet non fieri posse, nisi experimento vel rei, vel alterius rei similis, in eo saltem de quo agitur. At si quis dicat me id saltem posse cognoscere ex ideis in mente mea comprehensis, dum experior, me concipere EFG, quod voco A, respondeo me cum dico concipere E, vel concipere aliquid quod experior nihil involvere aliud, vel concipere aliquid adhuc compositum, quod a me confuse apprehenditur. Si experior E nihil involvere aliud seu per se concipi, tunc admitti potest ipsum esse possibile. Sed de tali nullae omnino fieri possunt propositiones, nisi identicae; alioqui falso dixi me experiri quod nihil aliud involvat. Si experior E involvere plura, jam ea rursus similiter tractanda sunt, quoties vero plura conjungo, quae non sunt per se concepta, opus est experimento, non tantum quod a me simul concipiantur in eodem subjecto, talis enim conceptus est confusus, sed quod revera extiterint in eodem subjecto.

(69) Itaque inter prima principia est, terminos quos in eodem subjecto existere deprehendimus non involvere contradictionem. Seu si A est B, et A est C, utique BC est possibile, seu non involvit contradictionem.

(70) Deus ex solis sui intellectus experimentis, sine ulla perceptione aliorum, judicat de rerum possibilitate.

(71) Quid dicendum de propositionibus A est existens, seu A existit. Ut si dicam de re existente A est B, idem est ac si dicam AB est existens, v.g. Petrus est abnegans, id est Petrus abnegans est existens.

(67) A necessary proposition, on the other hand, is one whose opposite is not possible, i.e. such that, if we assume its opposite, we reach a contradiction by means of resolution. Therefore, that proposition is necessary which can be demonstrated by means of identical propositions and definitions, without employing data of experience other than for showing that a term is possible.

(68) But it still remains to be examined how I know that I am proceeding correctly when I am defining something. For, if I say that A = EFG, then I must know not only that each of E, F, and G is possible in itself, but also that each is compatible with the others. It is clear, however, that this cannot be done except by experience, either of the thing itself or of another thing similar to it, in at least the respect under consideration. But if someone says that I can at least know this from the ideas contained in my mind, whilst I am experiencing that I conceive EFG, which I call A, then I reply that when I say that I conceive E, I can either conceive something which I know by experience to involve nothing else or conceive something which is again composite and is confusedly apprehended by me. If I experience E as involving nothing else, i.e. as being conceived through itself, then it can be admitted that it is possible. About such a thing, however, no propositions at all can be formed, with the exception of the identical ones; otherwise I stated falsely that I was experiencing something involving nothing else. If I am experiencing that E involves several things, these are again to be treated in a similar way; but as often as I join together several things which are not conceived through themselves, experience is needed, not only to secure that they are conceived by me at the same time in the same subject (for this concept, indeed, is confused), but to secure that they really exist in the same subject.

(69) Thus, one of the first principles is that terms which we learn exist in the same subject do not involve a contradiction. Or if A is B, and A is C, then BC too is possible, i.e. it does not involve a contradiction.

(70) God, on the basis only of the experiences of his own understanding, without any perception of anything else, makes judgements about the possibility of things.

(71) What is to be said about the propositions 'A is an existent' or 'A exists'? If I say of an existing thing, 'A is B', it is the same as if I were to say 'AB is an existent', e.g. 'Peter is a denier', i.e. 'Peter denying is an

Hic quaeritur quomodo in resolvendo procedendum sit, seu an terminus Petrus abnegans involvat existentiam; an vero Petrus existens involvat abnegationem, an omnino Petrus involvat et existentiam et abnegationem, quasi dicas Petrus est abnegans actu, seu abnegans existens, quod utique verum est. Et ita omnino dicendum est, et hoc discrimen est inter terminum individuum seu completum, et alium; nam si dicam aliquis homo est abnegans, homo non continet abnegationem, est enim terminus incompletus, nec homo continet omnia quae de eo dici possunt de quo ipse.

(72) Unde si sit BY, et terminus Y indefinitus quicunque sit superfluus; seu ut quidam Alexander Magnus, et Alexander Magnus sit idem, tunc B est *individuum*. ⟨Si sit terminus BA et B sit individuum, erit A superfluus, seu si BA = C, erit B = C⟩.

(73) Sed quaeritur quid significet τὸ Existens, utique enim existens est Ens seu possibile, et aliquid praeterea. Omnibus autem conceptis, non video quid aliud in Existente concipiatur, quam aliquis Entis gradus, quoniam variis Entibus applicari potest. Quanquam nolim dicere aliquid existere esse possibile seu Existentiam possibilem, haec enim nihil aliud est quam ipsa Essentia; nos autem Existentiam intelligimus actualem seu aliquid superadditum possibilitati sive Essentiae, ut eo sensu existentia possibilis futurum sit idem quod actualitas praescindens ab actualitate, quod absurdum est. Ajo igitur Existens esse Ens quod cum plurimis compatibile est; seu Ens maxime possibile. Itaque omnia coexistentia aeque possibilia sunt. Vel quod eodem redit, existens est quod intelligenti et potenti placet, sed ita praesupponitur ipsum Existere. Verum poterit saltem definiri, quod Existens est quod Menti alicui placeret, et alteri potentiori non displiceret si ponerentur existere mentes quaecunque. Itaque res eo redit, ut dicatur Existere quod Menti potentissimae non displiceret, si poneretur mens potentissima existere. Sed ut haec definitio applicari possit experimentis, sic potius definiendum est *Existit*, quod Menti alicui (existenti) placet (existenti, non debet adjici, si definitionem non simplicem propositionem quaerimus), nec Menti potentissimae (absolute) displicet. Placet autem menti potius id fieri

existent'. The question here is how we need to proceed with the resolution; i.e. whether Peter denying involves existence, or whether Peter existent involves the denial, or whether Peter plainly involves both existence and denial, as if you were to say 'Peter is actually a denier', i.e. an existent denier, which is certainly true. Thus, we should certainly speak in this way; and this is the difference between an individual or complete term and any other term; for, if I say 'some man is a denier', 'man' does not contain 'denial', as it is an incomplete term, nor does 'man' contain everything that can be said of that of which it can itself be said.

(72) Therefore, suppose that we have BY with Y any indefinite term, which is superfluous, i.e. so that 'some Alexander the Great' and 'Alexander the Great' are the same, then B is an *individual*. ⟨If there is a term BA and B is an individual, A will be superfluous, i.e. if BA = C, then B = C.⟩ *

(73) But the question is, however, what 'existent' means, for clearly an existent is a being, i.e. a possible and something else. All things considered, however, I do not see what is conceived in 'Existent' other than some degree of being, since it can be applied to various beings. I do not wish to say, however, that the existence of something is its being possible, i.e. its possible existence, because this is simply the essence itself. We understand existence, instead, as actual, i.e. as something added to possibility or essence, whereas in to the above sense, possible existence would be the same as actuality separated from actuality, which is absurd. I say, therefore, that an existing being is that which is compatible with the most things, i.e. a most possible being. Thus, all coexistent things are equally possible. Or, what comes to the same thing, that is existent which pleases one who is intelligent and powerful, but in this way it is presupposed that such a one exists. However, this definition at least can be given: that is existent which would please some mind and would not displease another more powerful mind, if minds of any kind were assumed to exist. Thus, it comes to this: something is said to exist that does not displease the most powerful mind, if it should be assumed that such a most powerful mind exists. But in order for this definition to be applicable to experience, we must rather give the following definition: that *Exists* which pleases some (existent) mind ('existent' must not be added if we seek a definition and not a simple proposition), and does not (absolutely) displease the most powerful mind. However, it is pleasing to a mind that what has a reason

quod habet rationem, quam quod non habet rationem, ita si plura sint A, B, C, D, et unum ex ipsis sit eligendum, et sint B, C, D per omnia similia, at solum A ab aliis sese aliqua re distinguat, Menti cuilibet hoc intelligenti placebit A. Idem est si saltem discrimen non appareat inter B, C, et D, appareat autem inter A et ipsa, et mens decreverit eligere, eliget A. Libere tamen eligit, quia potest adhuc inquirere, an non sit discrimen inter B, C, D.

(74) Omnes propositiones Existentiales, sunt verae quidem, sed non necessariae, nam non possunt demonstrari, nisi infinitis adhibitis, seu resolutione usque ad infinita facta, scilicet non nisi ex completa notione individui, quae infinita existentia involvit. Ut si dico Petrus abnegat, intelligendo de certo tempore, utique praesupponitur etiam illius temporis natura, quae utique involvit et omnia in illo tempore existentia. Si dicam infinite Petrus abnegat, abstrahendo a tempore; ut verum hoc sit, sive abnegarit, sive sit abnegaturus, tunc nihilominus saltem ex Petri notione res demonstranda est, at Petri notio est completa, adeoque infinita involvit, ideo nunquam perveniri potest ad perfectam demonstrationem, attamen semper magis magisque acceditur, ut differentia sit minor quavis data.

(75) Si, ut spero, possim concipere omnes propositiones instar terminorum, et omnes Hypotheticas instar Categoricarum, et universaliter tractare omnes, miram ea res in mea characteristica, et analysi notionum, promittit facilitatem, eritque inventum maximi momenti. Nimirum generaliter voco terminum falsum, qui in incomplexis est terminus impossibilis, vel saltem insignificans,[30] et qui in complexis est propositio impossibilis, vel saltem propositio quae probari non potest. Itaque manet analogia.[31] Itaque per A intelligo vel terminum incomplexum, vel propositionem; vel collectionem, vel collectionum collectionem, etc. Ut generaliter terminus verus sit, qui perfecte intelligi potest.

(76) Praeter Ens adhibebimus etiam Entia, unde prodit totum et pars. Generaliter si A non est B et B non est A, et primitiva est haec: A est L et B est L idem esse quod C est L, dicitur C totum, A (aut B)

[30] NB.   [31] NB.

be made, rather than what does not have a reason. Thus, if there are several things A, B, C, D, and we have to choose one of them, and if B, C, D are similar in all respects, whereas A alone is distinguished from the rest in some way, then A will please any mind which understands this. The same holds even if a distinction does not appear between B, C, and D, but does appear between them and A, and the mind has decided to choose: it will choose A. It chooses freely, for it can still inquire whether there is not a distinction between B, C, and D.

(74) All existential propositions are certainly true, but not necessarily so, for they cannot be demonstrated except by recourse to infinitely many things, i.e. by a resolution which will go through infinitely many facts, that is, only from the complete notion of an individual, which involves infinitely many existing things. Thus, if I say 'Peter denies', understanding this as referring to a certain time, undoubtedly the nature of that time is presupposed too, which also involves everything existent at that time. If I say 'Peter denies' indefinitely, abstracting from time, so that it is true whether he has denied or will deny, then it must nevertheless be demonstrated from the concept of Peter at least. But the concept of Peter is complete and what is more it involves infinite things; therefore, we can never arrive at a perfect demonstration, even though we always approach it more and more, so that the difference is less than any given.

(75) If, as I hope, I can conceive all propositions as terms, and all hypotheticals as categoricals, and if I can treat them all in the same way, this promises a wonderful facility in my characteristic art and analysis of concepts, and it will be a discovery of the greatest importance. Clearly, in general, I call a term 'false' which in the case of incomplex terms is an impossible term, or at least a meaningless one,[i] and which in complexes is an impossible proposition, or at least a proposition which cannot be proved.[ii] Thus, the analogy remains. So, by 'A' I understand either an incomplex term or a proposition or a collection or a collection of collections, etc., so that, in general, a term is true which can be perfectly understood.

(76) Besides 'being', we shall also employ 'beings', from which whole and parts arise. In general, if A is not B and B is not A, and the proposition 'A is L and B is L is the same as C is L' is primitive, then C is called a

---

[i] NB. [ii] NB.

pars.[32] Dubitari potest an et quatenus C sit unum Ens reale, an non semper ex pluribus resultet novum Ens, etiam dissitis, et quandonam resultet vel non.

(76[⟨bis⟩]) Non-A est non $\overline{AB}$, seu non-A = Y non $[\overline{AB}]$.[33] Omnis non homo est non: $\overline{homo\ rationalis}$, sequitur ex 77.

(77) Generaliter A esse B idem est quod non-B est non-A. Unde demonstratur praecedens: nam AB est A. Ergo non-A est non $[\overline{AB}]$.[34]

(78) A = B et non-A = non-B coincidunt.

(79) At si A sit B, non sequitur non-A esse non-B, seu si homo sit animal, non sequitur non hominem esse non animal. Itaque licet pro A substitui possit B, non ideo tamen pro non-A licet substituere non-B, nisi vicissim pro B substitui possit A.

(80) Videndum an infinitis possit careri, sane non-A videtur idem esse quod is qui non est A, seu subjectum propositionis negativae cujus praedicatum est A, seu omnis qui non est A. Itaque si $\overline{Y}$ non est A, erit $\overline{Y}$ = non-A, seu $\overline{Y}$ non = AX idem est quod $\overline{Y}$ = [X] non-A.[35]

(81) $\overline{Y}$ seu Y indefinita cum lineola mihi significat quilibet, Y est unum incertum, $\overline{Y}$ est quodlibet.

(82) Nimirum et sic dici poterit: B non est A idem esse quod, B est non-A, unde B non = AY idem esse quod B = Y non-A.

(83) Generaliter A esse B, idem est quod A = AB, inde enim manifestum est B contineri in A, idemque est homo, et homo animal. Notavi hoc jam supra ad marginem articuli 16, et quanquam inde fieri ⟨videatur homo est rationale animal animal, tamen animal animal idem est quod animal, ut notavi supra articulo 18⟩.

(84) Hinc si propositio A est B dicatur esse falsa seu negetur, utique hoc est dicere A non = AB hoc est quoddam A non est B.

---

[32] Continuum cum partes indefinitae.
Numerus oritur si consideratur tantum plura esse Entia, non qualia.
[33] [L: AB].
[34] [L: non B]. [This marginal remark follows:] Hoc videndum an possit demonstrari. Demonstratum est infra 95 et 99.
[35] [The manuscript has: '$\overline{Y}$ = non-A', and the Academy edition corrects it to '$\overline{Y}$: non-A', but I do not see a reason for that correction].

'whole' and A (or B) a 'part'.^{y} It is possible to doubt whether and to what extent C is one real being; whether a new being does not always results from several, even when they are scattered, and when it does or does not result.

(76[bis]) Non-A is not [$\overline{AB}$], or non-A = Y not [$\overline{AB}$]. Every non-man is a non-rational *man*. This follows from 77.

(77) In general, that A is B is the same as 'non-B is non-A', from which we demonstrate the preceding proposition, because AB is A. Therefore, non-A is not [$\overline{AB}$].^{w}

(78) A = B and non-A = non-B coincide.

(79) But if A is B, it does not follow that non-A is non-B, i.e. if man is an animal, it does not follow that a non-man is a non-animal. Thus, although B can be substituted for A, we cannot substitute non-B for non-A, unless A in turn can be substituted for B.

(80) We must see whether it is possible to avoid infinite terms.* Certainly, 'non-A' seems to be the same as 'the one who is not A', i.e. the subject of a negative proposition whose predicate is A, or 'everyone who is not A'. Thus, if $\overline{Y}$ is not A, then $\overline{Y}$ = non-A, i.e. '$\overline{Y}$ not = AX' is the same as '$\overline{Y}$ = [X] non-A'.

(81) I take '$\overline{Y}$,' i.e. the indefinite 'Y' with a little line on it, to signify 'anyone'; 'Y' is one indeterminate thing; '$\overline{Y}$' is anything whatever.

(82) Of course, it will also be possible to say that 'B is not A' is the same as 'B is non-A'; therefore, 'B not = AY' is the same as 'B = Y non-A'.*

(83) In general, that A is B is the same as A = AB: from this, indeed, it is clear that B is contained in A and that 'man' and 'man animal' are the same. I have already noted this in the margin of article 16 above; and even though ⟨'man is a rational animal animal' seems to follow from this, yet 'animal animal' is the same as 'animal', as I remarked above in article 18⟩.

(84) Hence if the proposition 'A is B' is said to be false, i.e. is denied, this clearly amounts to saying that A not = AB, i.e. some A is not B.

---

^{y} There is a continuum when the parts are indefinite.
A number arises if we consider only that beings are several, but not of what kind they are.
^{w} We must see whether this can be demonstrated.
It has been demonstrated below, 95 and 99.

(85) A esse non-B idem est ac dicere A = A $\overline{\text{non-B}}$, patet ex 83. Si dicas 'A = A non-B est falsa', seu 'A [non = A]^{36} non-B', significat 'quoddam A est B'.

(86) Rursus non-B idem est quod is qui non est B, seu genus cujus species sunt A, C, D, etc., posito A non esse B, C non esse B, D non esse B.

(87) Itaque Nullum A esse B idem est quod A esse non-B, seu quodlibet A esse unum ex iis quae non sunt B. Seu [A$\overline{\text{Y}}$ non = AB$\overline{\text{Y}}$],^{37} idem est quod A = A non-B. Habemus igitur transitum inter infinitas affirmativas, et negativas.

(88) Ut obiter dicam, generaliter A esse AB, idem est quod A coincidere cum AB (seu si propositio A est AB est vera, erit reciproca). Hoc ita demonstro: A est AB ex hypothesi, id est (per 83) A = AAB, id est (per 18) A = AB. Idem sic: A est AB (ex hypothesi) et AB est A (per 38). Ergo (per 30) A = AB. Hae duae demonstrationes inter se comparentur, aut enim in idem desinent, aut dabunt demonstrationem alicujus propositionis sine probatione assumtae.^{38}

(89) Consideremus particularem affirmativam quoddam animal est homo, BY = AZ. Ea etiam potest in hanc mutari BY = ABY seu dici potest quoddam animal esse hominem, idem esse quod, animal quoddam esse hominem-animal. Patet ex 83. Nihil refert enim quod Y incerta est, quaecunque enim illa sit, fingatur nosci, et adesse, tunc utique locum haberet ratiocinatio.

(90) Caeterum etsi hoc modo in Praedicato vitari semper possit indefinita Y, non tamen potest vitari in subjecto, et praestat praedicato etiam relinqui, ob inversionem manifestiorem. Et omnino quia non prorsus eliminari possunt indefinitae, praestat eas relinqui.^{39}

(91) Si A est B tunc A non est non-B. Esto verum A esse non-B, si quidem fieri potest, jam A est B ex hypothesi. Ergo A est B non-B, quod est absurdum. Adde infra [100^{40}].^{41}

---

^{36} [L: A non-A = non-B]. ^{37} [L: AY non = ABY]. ^{38} NB.
^{39} Imo puto posse eliminari. ^{40} [L: 99].
^{41} Hic ratiocinandi modus, seu ducendi ad absurdum, jam in praecedentibus est opinor stabilitus.

(85) That A is not B is the same as saying 'A = A $\overline{\text{non-B}}$'; this is evident from 83. If you say 'A = A non-B is false', i.e. 'A [non = A] non-B', then this means 'some A is B'.

(86) Again 'non-B' is the same as 'he who is not B' i.e. a genus whose species are A, C, D, etc., granted that A is not B, C is not B, and D is not B.

(87) Thus, that no A is B is the same as that A is non-B, i.e. that any A is one of those things which are not B; or [A$\overline{\text{Y}}$ non = AB$\overline{\text{Y}}$] is the same as A = A non-B. Therefore, we have a passage between infinite affirmative and negative propositions.*

(88) Incidentally: in general, that A is AB is the same as that A coincides with AB (i.e. if the proposition 'A is AB' is true, it will be reciprocal). I demonstrate this as follows: A is AB *ex hypothesi*, i.e. (by 83) A = AAB, i.e. (by 18) A = AB. Likewise: A is AB (*ex hypothesi*) and 'AB is A' (by 38). Therefore, (by 30) A = AB. These two demonstrations should be compared with each other, for either they will end in the same thing, or they will give a demonstration of some proposition that has been assumed without proof.ˣ

(89) Let us consider the particular affirmative proposition 'some animal is a man', BY = AZ. It also can be changed into this: BY = ABY, that is, it can be said that that some animal is a man is the same as that some animal is a man-animal. This is evident from 83. For it does not matter that 'Y' is uncertain, because whatever it may be, it may be imagined to be known and to be present; then the reasoning would be successfully performed.*

(90) However, although in this way the indefinite 'Y' can always be avoided in the predicate, it still cannot be avoided in the subject; thus it is better to leave it in the predicate as well, to make the inversion more evident. In general, since indefinite letters cannot be absolutely eliminated, it is better to leave them.ʸ

(91) If A is B, then A is not non-B. Let it be true that A is non-B, if in fact it is possible. Now, A is B *ex hypothesi*; therefore, A is B non-B, which is absurd. Add no. [100] below.ᶻ

---

* NB.   ʸ Yet I think that they can be eliminated.
ᶻ I think that this way of reasoning, i.e. of reducing to the absurd, has already been established.

(92) Non valet consequentia: Si A non est non-B, tunc A est B, seu omne animal esse non hominem falsum est, quidem; sed tamen hinc non sequitur omne animal esse hominem.

(93) Si A est B, non-B est non-A. Falsum esto si fieri potest non-B esse non-A, seu non-B non esse A, verum erit non-B esse A. Ergo quoddam A est non-B. Ergo falsum est omne A esse B, contra Hyp.

(94) Si non-B est non-A, A est B. Falsum esto si fieri potest A esse B, ergo A erit non-B. Ergo quoddam non-B erit A (per conversionem). Ergo falsum est quoddam non-B esse non-A (per 91). Ergo multo magis falsum est omne non-B esse non-A, contra hypothesin.

(95) A esse B idem est quod non-B esse non-A, patet ex 93, 94, juncto 30. Videndum an non propositio 95 demonstrari possit per se, sine 93 et 94. Hoc praestitum articulo [99].⁴²

(96) Non-non-A = A.⁴³

(97) Nullum A est B idem est quod A est non-B (per 87).

⟨(98) Omne A est B idem est quod Nullum A est non-B, seu quoddam A non esse non-B. Patet ex 97 vel 87, tantum pro B ponendo non-B et pro non-B ponendo non-non-B seu B.⟩

(99) A est B idem quod A est non-non-B (per 96) et hoc idem (per 87) quod Nullum A est non-B id est nullum non-B est A (per conversionem universalis negativae) id est (per 87) Omne non-B est non-A = A est B. Quod erat dem.

(100) Si A est B, sequitur A non esse non-B, seu falsum esse Omne A esse non-B. Nam si A est B, utique nullum A est non-B, seu falsum est quoddam A esse non-B (per 87). Ergo (per 101) multo magis falsum est Omne A esse non-B. Adde 91.

(101) Si falsum est aliquod A esse B, falsum est omne A est B seu quod idem est aliquod A non est B. Ergo omne A non est B. Nam ponatur si fieri potest omne A esse B. Ergo quoddam A est B (per 29). Sed hoc est contra hypothesin, adeoque falsum, ergo et falsum prius.

(102) Si A est B et A est C, idem hoc est quod A est BC.

(103) Hinc si A est non-B et A est non-C idem hoc est quod A est non-B non-C.

---

⁴² [L: 98].
⁴³ [On the margin of a first (then deleted) version of (96), Leibniz has written:] Nullum non-A, idem est quod Solum A [No non-A is the same as A alone].

(92) The consequence: 'if A is not non-B, then A is B' is invalid; that is, it is clearly false that every animal is a non-man, but nevertheless it does not follow from this that every animal is a man.*

(93) If A is B, non-B is non-A. Let it be false (if possible) that non-B is non-A, i.e. that non-B is not A: then it will be true that non-B is A.* Therefore, some A is non-B. Therefore, it is false that every A is B, contrary to the hypothesis.ªª

(94) If non-B is non-A, A is B. Let it be false, if possible, that A is B; therefore, A will be non-B.* Therefore, some non-B will be A (by conversion). Therefore, it is false that some non-B is non-A (by 91). Therefore, it is much more false that every non-B is non-A, contrary to the hypothesis.

(95) That A is B is the same as non-B is non-A is evident from 93 and 94, with the addition of 30. We must see whether or not proposition 95 can be demonstrated by itself, without 93 and 94. This is shown by article [99].

(96) Non-non-A = A.

(97) 'No A is B' is the same as 'A is non-B' (by 87).

⟨(98) 'Every A is B' is the same as 'No A is non-B', i.e. that some A is not non-B. This is evident from 97 or 87 by simply putting 'non-B' for 'B' and 'non-non-B', i.e. 'B', for 'non-B'.⟩

(99) 'A is B' is the same as 'A is non-non-B' (by 96), and this is the same (by 87) as 'No A is non-B', i.e. 'No non-B is A' (by conversion of the universal negative). That is (by 87) Every non-B is non-A = A is B.Q.E.D.

(100) If A is B, it follows that A is not non-B, i.e. that it is false that every A is non-B. For if A is B, certainly no A is non-B, i.e. it is false that some A is non-B (by 87). Therefore, (by 101) much more false is that every A is non-B. Add 91.

(101) If it is false that some A is B, it is false that 'every A is B' or, what is the same, 'some A is not B'.* Therefore, every A is not B. For let it be posited, if possible, that every A is B. Then, some A is B (by 29). But this is contrary to the hypothesis, and for this reason is false. Therefore, the former is false as well.

(102) If A is B and A is C, this is the same as 'A is BC'.

(103) Hence, if A is non-B and A is non-C, this is the same as 'A is non-B non-C'.

(104) Non-B esse non $\overline{BC}$ demonstratum est 76[bis]. Sed non semper [non-$\overline{BC}$]^{44} est non-B. Excogitandus esset modus propositionis formalis, seu generalis, quasi dicerem: falsum est omne negativum compositum esse negativum simplex seu non $\overline{\overline{Y}}$ $\overline{\overline{X}}$ non = non-$\overline{\overline{Y}}$,^{45} ita ut $\overline{\overline{Y}}$ et $\overline{\overline{X}}$ significent quaslibet similiter se habentes.

(105) Si A est non $\overline{BC}$ non ideo sequitur vel A esse non-B, vel A esse non-C, ⟨potest enim fieri ut B sit = LM et C = NP, et ut A sit [non $\overline{LN}$],^{46} quo facto A erit non $\overline{LMNP}$ seu non $\overline{BC}$⟩, interim hinc sequitur falsum esse simul A esse B et A esse C seu A esse BC. Patet ex 91 vel [100].^{47}

(106) Patet ex his *non* a sua litera vel formula cui praefigitur in calculo divelli minime debere.

(107) Omnis complicatio propositionum ita generaliter repraesentari potest $\overline{\overline{AB}CD}$, etc. vocare possumus $\overline{AB}$ = L, $\overline{LC}$ = M, $\overline{MD}$ = N ponendo aliqua horum similiter posse resolvi ut L vel M vel N, et ea in quae ipsa resolvuntur, rursus ita fortasse posse resolvi, pro re nata. Lineola autem supra ducta ut $\overline{AB}$ significare potest affirmationem vel negationem aut potius coincidentiam vel incoincidentiam. Poteritque lineola notam quandam habere tam in medio quam in extremis, in medio ut significetur modus propositionis, utrum sit affirmativa an negativa, etc., extremum autem quo respicitur A poterit notam habere qua designetur utrum A sit terminus universalis an particularis, etc. similiter idem designabit pro B lineola quae respicit B. Et si sit

$$\begin{array}{ccc} 4 & & 5\ 6 \\ \underline{1} & \underline{2} & \underline{3} \\ A & B & C \end{array}$$

locus 1 designabit quantitatem vel qualitatem etc. secundum [quam]^{48} hic adhibetur terminus A seu modum adhibendi termini A, et locus 2 naturam propositionis AB, locus 3 modum termini B. Locus 4 modum adhibendi τοῦ AB seu L. Locus 5 naturam propositionis $\overline{ABC}$ seu $\overline{LC}$,

^{44} [L: non-BC]. ^{45} NB. ^{46} [L: non-LN]. ^{47} [L: 99].
^{48} [L: quem].

(104) That non-B is non-$\overline{BC}$ has been demonstrated in 76[bis]. But it is not always the case that [non-$\overline{BC}$] is non-B. We should have found a formal or general way of representing a proposition, as if I were saying: 'it is false that every composite negative is a simple negative', i.e. non $\overline{\overline{Y}}$ $\overline{\overline{X}}$ non = non-$\overline{\overline{Y}}$, [bb] so that '$\overline{\overline{Y}}$' and '$\overline{\overline{X}}$' signify any propositions standing in a similar relation to each other.

(105) If A is not $\overline{BC}$, it does not therefore follow either that A is non-B or that A is non-C, ⟨for it can happen, indeed, that B is = LM and C = NP, and that A is [⟨not $\overline{LN}$⟩], in which case A will be not $\overline{LMNP}$, i.e. not $\overline{BC}$. From this, instead, it follows that it is false that at the same time A is B and A is C or that A is BC. This is evident from 91 or [100].

(106) It is clear from this that '*non*' ought to be separated as little as possible from the letter or formula to which it is prefixed in the calculus.

(107)* Every complex of propositions can thus be represented generally as '$\overline{ABCD}$', etc. We can say that $\overline{AB}$ = L, $\overline{LC}$ = M, $\overline{MD}$ = N, assuming that some of these, such as L or M or N, can be analysed in a similar way and that those into which they themselves are analysed, can perhaps be analysed in their turn, according to the circumstances. A little line drawn above, however, such as '$\overline{AB}$' can signify affirmation or negation, or rather coincidence or non-coincidence; and it can have some mark in the middle and at the ends as well: in the middle to signify the mode of the proposition, whether it is affirmative or negative, etc., whilst the extreme which is over 'A' can have a mark which will indicate whether A is a universal or a particular term, etc., and analogously the extreme which is over B will indicate the same for B. And if we have:

$$\begin{array}{ccc} 4 & & 5 \ 6 \\ \hline 1 & 2 & 3 \\ \hline A & B & C \end{array}$$

place 1 will indicate the quantity or quality, etc. according to which the term A is employed here, i.e. according to the way of using the term A; place 2 will designate the nature of proposition AB; place 3 the mode of the term B; place 4 the mode of employing AB, i.e. L; place 5 the nature of the proposition $\overline{ABC}$, i.e. $\overline{LC}$;

[bb] NB.

locus [6][49] modum termini [C].[50] Posset in numeris observari talis ordo, ut semper incipiatur a maxime subdivisis seu ab infimo subdivisionis gradu seu a terminis ad incomplexa propioribus ut si sit

```
[1] 3       14        15
 10   11        12
   7  8   9                    ⊙
   1 2 3       4 5 6
 A  B  C    D     E  F
```

Unde intelligi potest quam miris modis terminorum relationes et denominationes variari possint tam ab ordine si respicias solam dispositionem numerorum, quam a valore cujusque numeri, si vel solius quantitatis et qualitatis habeatur locus.

(108) Omnis terminus etiam incomplexus potest haberi pro propositione, quasi ipsi adjectum esset τὸ hoc Ens, ut Homo perinde sumi potest ac si diceretur Homo idem est quod est hoc Ens, scilicet id ipsum quod est, vel potius generalius, perinde erit ac si adjectum esset τὸ verum, ut: Homo est verum, $\overline{\text{Homo est animal}}$ est hoc verum et τὸ hoc verum facit hoc loco officium quod unitas in Arithmetica, ad supplenda loca seu dimensiones. Si scilicet ponatur quodlibet quod cum aliquo copulatur tot modis esse subdivisum quo id cum quo copulatur, ne terminus nisi aeque complexo vel incomplexo jungi ponatur, verum seu Unitas scribatur V, ex ⊙ fiet ☽, ubi loca sunt suppleta, dici enim potest $\overline{\text{A esse idem quod hoc verum}}$, esse idem quod, hoc verum, est hoc verum: sed notandum ipsum V suppletum ubique debere mutari: A = A verum seu [A est hoc verum].

```
                        ☽
  43                    44                                        45
 37         38          39  40          41                        42
 25    26 27  28 29     30  31    32    33 34      35             36
 1 2  3  4 5 6  7 8 9  10   11    12  13 14 15  16 17 18  19 20 21  22   23 24
 A    V  V   V  B V  C          D  E     V  V     V  F     V  V            V
```

[49] [L: 5].   [50] [L: B].

place [6] the mode of employing the term [C].* An order could be observed in numbers such that we always begin from the most subdivided, i.e. from the lowest degree of subdivision, that is, from terms which are nearer to the incomplex ones, as if there were, for instance:

```
 13           14        15
 10    11         12
    7  8  9                    ⊙
       1 2 3      4 5 6
    A  B C  D     E   F
```

From this it can be understood in what extraordinary ways the relations and denominations of terms can be varied, as much by the order, if you consider only the disposition of numbers, as by the value of each number, if only quantity and quality are taken into account.

(108) Every term, even an incomplex one, can be considered a proposition, as if 'this being' were added to it, just as 'man' can be understood as if it were said 'man is the same as what this being is', namely 'the very same thing that it is', or rather more generally, precisely as if 'true' were added, e.g. 'man is a true thing', 'man is an animal is this true thing'. Here, 'this true thing' plays the same role as unity in arithmetic, to supply places or dimensions. That is, if we suppose that anything which is joined with another thing is subdivided in as many ways as that with which it is joined, so that a term is supposed to be joined only to one which is equally complex or incomplex, and 'true thing' or 'unity' is written 'V', then we shall get, instead of the preceding table marked with ⊙, that marked with ☽ where the places have been supplied. For we can say that 'A is the same as this true thing' is the same as 'this true thing is this true thing'. But it should be noted that the 'V' which has been supplied must be changed everywhere: 'A = A is a true thing' i.e. ⟨'A is this true thing'⟩*

```
                           ☽
 43                        44                              45
 37          38            39  40            41            42
 25    26 27  28 29         30  31   32  33  34     35      36
 1 2  3  4 5 6 7 8 9 10  11  12  13 14 15 16 17 18 19 20 21 22  23 24
 A    V  V V  B  V  C        D   E   V  V     V  F   V  V        V
```

(109) Quemadmodum autem quilibet terminus concipi potest instar propositionis, ut explicuimus, ita et quaelibet propositio concipi potest instar Termini, ut Hominem esse animal est verum, est propositio, est tale quid, est causa, est ratio, etc. Quae serviunt ad universalissimas condendas enuntiationes de his complicationibus.

(110) Possunt etiam novi Termini reflexivi condi, qui similiter tractari possunt ut directi, ut subjectum propositionis talis, tale..., potest appellari aliquo nomine. Et videndum quomodo hae ipsae denominationes rursus inter se per literas explicari possint, ut si subjectum propositionis universalis affirmativae sit praedicatum alterius propositionis affirmativae, cujus subjectum est praedicatum prioris, subjectum dicitur esse idem cum praedicato ejusdem propositionis. Si quis autem velit rigorose rem enuntiari ad morem communem logicorum ⟨aut etiam hominum vulgo loquentium⟩ in propositionibus satis difficultatis inveniet, ut si dicere velit subjectum propositionis universalis affirmativae, cujus praedicatum est subjectum propositionis universalis affirmativae in qua [praedicatum est subjectum][51] praecedentis propositionis, est idem cum praedicato dictae propositionis cujus est subjectum. Ac ne sic quidem relativum, dictae, vel praecedentis, potest evitari, quanto satius, brevius, clariusque dicemus si A est B et B est A, A est idem cum B. Cujus etiam demonstratio facile dari potest, quemadmodum supra a nobis data est; adhibitis scilicet literis. At verbis haud dubie foret satis perplexa, et opus foret peculiarem adhibere curam in illis recte disponendis. Nam si recte constituta essent, credo idem praestarent, licet nesciam an pari claritate, similiter et consequentiae ex literis facile ducuntur, ut statim hic patet ut A diximus esse idem ipsi A, ita et B posse dici idem ipsi B, quod non aeque videtur facile ex verbis apparere.

(111) Notandum est posse etiam de tota resolutionis serie generalia quaedam excogitari circa processum ejus, etiamsi continuaretur resolutio in infinitum, et circa haec utique excogitari possent verba apta

---

[51] [L: subjectum est praedicatum].

(109) Just as any term can be conceived as a proposition, as we explained, so also any proposition can be conceived as a term, as 'man's being an animal is true, is a proposition, is some such a thing, is a cause, is a reason', etc. And these transformations are useful for building the most universal statements concerning these combinations.

(110) New reflexive terms can also be composed which can be treated similarly to direct ones: 'the subject of such a proposition', 'such a thing…', for example, can be called by some name.* And we must see how these denominations can again be mutually explained by means of letters. If, for example, the subject of a universal affirmative proposition is the predicate of another affirmative proposition, whose subject is the predicate of the former, the subject is said to be identical with the predicate of the same proposition. If, however, one wants this to be stated rigorously according to the common practice of the logicians, ⟨or even of men as they ordinarily speak⟩, one will meet with considerable difficulties when dealing with propositions. This is the case, for instance, if we want to say: 'the subject of a universal affirmative proposition, whose predicate is the subject of a universal affirmative proposition, in which [the predicate is the subject] of the preceding proposition, is the same as the predicate of the proposition mentioned of which it is the subject'. Even so, however, we cannot avoid the relative expression 'mentioned', or 'the preceding'; how much more adequately, briefly and clearly, then, shall we say, 'if A is B and B is A, A is the same as B'. And of this a demonstration can easily be given, as we have given it above i.e. with the use of letters. But if words were employed, the demonstration would doubtless be rather convoluted and particular care would be needed to arrange them in the right order. For if they were arranged in the right order, I believe that they would give the same result, although I do not know whether they would do so with the same clarity. Analogously, consequences can be easily drawn from letters; thus, it is immediately evident here that, as we have said that A is the same as A, so too can B be said the same as B, which does not seem to be made clear so easily from words.

(111) It is worth noting that even in the case of the entire series of a resolution certain general features can be discovered about its progress, even though the resolution is endlessly continued; and concerning these, too, certain appropriate reflexive words can be discovered, or even some

reflexiva, vel etiam literae quaedam generales ut $\overline{\overline{Y}}$, sed in progressu clarius apparebit, quid horum praestet.

(112) Videndum an non alio nonnihil sensu sumatur Y cum dicatur AY est B hoc est quoddam A est B, quam cum negatur ullum A esse B, ita ut non tantum negetur quoddam A esse B seu incertum hoc A esse B, sed et quodcunque ex incertis A, ut proinde cum dicitur nullum A esse B, sensus sit negari $A\overline{Y}$ esse B, nempe $\overline{Y}$ est Y, seu quodcunque Y continebit hoc Y. Itaque cum dico quoddam A est B, dico hoc quoddam A est B, si nego quoddam A esse B, seu hoc quoddam A esse B, tantum videor particularem negativam dicere. At cum nego quodcunque A esse [⟨B⟩], seu non tantum hoc, sed et hoc et hoc A esse B, tunc nego $\overline{YA}$ esse B. Unde etiam in loquendo negare quoddam A esse B, seu dicere quoddam A non est B, non videtur sonare nullum A esse B, et similiter dicere Omne A non est B, non videtur sonare negationem quod omne A sit B; sed dici de quolibet A, quod non sit B. Pro prioribus tamen stat, quod negatio Universalis affirmativae, utique est particularis negativa.[52] Itaque negatio particularis affirmativae non potest etiam esse particularis negativa (neque enim negatio particularis affirmativae et universalis affirmativae potest esse idem), superest ergo, ut sit universalis negativa; neque enim aliud esse potest.

(113) Res utiliter exhibebitur figuris. A est B seu A coincidit cuidam B, seu A coincidit AB

⟨Lineola perpendicularis significat limites ultra quos non possunt et intra quos possunt extendi termini salva propositione seu habitudine.

---

[52] Univ. Aff. A aequatur B cum aliquo addito. Univ. Neg. Negatur.

general letters as '$\overline{\overline{Y}}$'. But which of these is best will become clearer as we proceed further.

(112) When we say 'AY is B', i.e. 'some A is B', it must be seen whether Y is not taken in some other sense than when we deny 'any A is B', in such a way that not only is it denied that some A is B, i.e. that this indeterminate A is B, but also that any A whatever from among the indeterminate A, is B, so that when we say that no A is B, the sense is that it is denied that A$\overline{Y}$ is B, given that $\overline{Y}$ is Y, i.e. any Y whatever will contain this Y. And so when I say 'some A is B', I am saying 'this particular A is B'; if I deny that some A is B, i.e. that this particular A is B, I seem only to be asserting a particular negative. But when I deny that any A whatever <u>is B</u>, i.e. that not only this, but also this and this A is B, then I deny that $\overline{YA}$ is B. So even in speaking, to deny that some A is B, or to say 'some A is not B', does not seem to signify 'no A is B', and, similarly, to say 'every A is not B' does not seem to signify the negation of 'every A is B'; rather, it is said of any A whatever that is not B. According to what we have stated above, however, the negation of a universal affirmative is undoubtedly a particular negative. Therefore, the negation of a particular affirmative cannot be, in its turn, a particular negative (for the negation of a particular affirmative and that of a universal negative, cannot be the same). What remains, therefore, is that it is a universal negative; for it cannot be anything else. *

(113) This will usefully be shown by means of figures. 'A is B', i.e. A coincides with some B, i.e. A coincides with AB:

⟨A small perpendicular line* signifies the limits beyond which terms cannot, and within which they can, be extended without affecting the proposition, i.e. the arrangement of the terms.⟩

---

* Universal Affirmative proposition: A is the same as B with something added. Universal Negative proposition: It is denied.

Ut lineola perpendicularis significat maximum, ita duplex linea horizontalis significat minimum seu quod detrahi non potest salva habitudine, duplex linea non videtur in subjecto necessaria, sed tantum in praedicato, subjectum enim sumo pro arbitrio. Pro duplici malo fortiorem. Ut quando linea proxime sub linea ducitur intelligatur unus terminus componi licet etiam semper intelligi possit unus respectu magis distantium linearum adhuc inferius ductarum.

(114) Quoddam A est B, seu quoddam A coincidit cuidam B.

(115) Hinc A = A. Nimirum generaliter fingendum est, quasi lineae horizonti parallelae, quarum una ducta est sub alia distinctionis causa, [ductae][53] essent una super alia.

116) AB = BY, ubi per Y intelligo quicquid est in tota linea B quod cadit sub A.
117) A = BY, idem est quod A = BA.
118) A = BY. Ergo BY = AY.
119) A = BY et B = AY idem est quod A = B

AB in genere

Haec omnia ex figurae inspectione patent

(120) Negatio hujus: quoddam A esse B seu cum negatur quoddam A coincidere cuidam B, sic exprimetur:

(121) Sed negatio hujus: Omne A est B sic exprimetur:

___
[53] [L: ducta].

Just as a small perpendicular line signifies the maximum, so a double horizontal line signifies the minimum, i.e. what cannot be subtracted without affecting the arrangement of the terms. The double line does not seem to be necessary in the subject, but only in the predicate, for I assume the subject arbitrarily. Instead of a double line, I prefer a heavier one. So, when a line is drawn very closely below another, it is understood that one term is composed, though it may also always be understood as one in the case of more distant lines drawn still farther down.

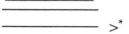

$>^*$

(114) 'Some A is B', i.e. some A coincides with some B.

A

B

(115) Hence, A = A. In general, the parallel horizontal lines, of which one is drawn below the other in order to distinguish them, should be imagined as if they were drawn one upon the other.

116) AB = BY, where by 'Y' I mean whatever there is in the entire line B, which falls under A.
117) A = BY is the same as A = BA.
118) A = BY. Therefore BY = AY.
119) A = BY and B = AY is the same as A = B

AB in general

A

B

All these things are manifest from inspection of the figure.

(120) The negation of this: 'some A are B', i.e. when it is denied that some A coincide with some B, will be expressed thus:

A

B

(121) But the negation of this: 'every A is B' will be expressed thus:

A

B

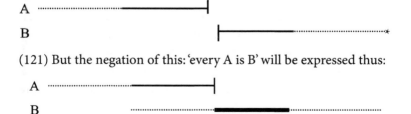

(122) Potest et alia consideratio institui, ut genus non ponatur esse pars speciei, ut paulo ante fecimus, quia generis notio est pars (vel saltem inclusum) notionis speciei; sed ut contra potius species sit pars generis, quia individua speciei sunt pars (vel saltem inclusum) individuorum generis.

(123) Itaque Omne A est B sic repraesentabitur

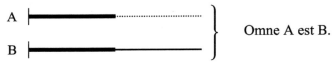

$$\left. \begin{array}{l} A \\ \\ B \end{array} \right\} \quad \text{Omne A est B.}$$

quae repraesentatio est inversa prioris. Eodem modo repraesentatio particularis negativae est inversa prioris. Sed particularis affirmativa et Universalis negativa eodem modo repraesentantur ut ante, quia nihil refert utrum praeponas aut postponas, itaque generaliter dici potest priorem repraesentationem a posteriore in eo saltem differre, quod lineae in figura transponuntur.

(124) Est et alia repraesentatio propositionum per numeros. Nempe pro terminis ponendo numeros, Universalis affirmativa seu A est B significat: A (vel saltem quadratum ipsius A aut cubus) dividi potest per B. Nam A, A^2, A^3, hic habentur pro iisdem.

(125) Particularis affirmativa, quoddam A est B, significat A multiplicatum per B seu AB dividi posse per B. Intellige scilicet ⟨AB semper dividi posse per [B][54]⟩, nisi in AB destruatur [B],[55] si verbi gratia A significaret $\frac{C}{B}$ et C non posset dividi per B.

(126) Particularis negativa est, falsum esse dividi A posse per B, licet forte AB dividi possit per B.[56]

(127) Universalis negativa est falsum esse AB dividi posse per B, cujus nulla alia causa est, quam quod A continet $\frac{1}{B}$. Itaque proprie universalis negativa est si A continet non-B,[57] unde per consequentiam colligitur Universalem negativam esse oppositam particulari affirmativae, nempe si A dividitur per B non potest fieri, ut A per B multiplicetur.

⟨Distinguenda negatio a divisione, divisione fit omissio alicujus termini, sed non ideo negatio nisi quod revera in infinitis, quod non inest

---

[54] [L: A].   [55] [L: A].
[56] Omnia per numeros demonstrari possunt, si modo notetur.
[57] NB.

(122) A different consideration can be established: that we do not suppose the genus to be a part of the species, as we did a little earlier on the grounds that the notion of the genus is a part of (or at least something included in) the notion of the species; but that, on the contrary, the species is a part of the genus, because the individuals of the species are a part of (or at least included among) the individuals of the genus.*

(123) So 'every A is B' will be represented thus

And this representation is the inverse of the preceding one.* In the same way the representation of a particular negative proposition is the inverse of the preceding one. But a particular affirmative proposition and a universal negative one are represented in the same way as before, since it does not matter whether you put a line in front or behind, and so it can be said generally that the former representation differs from the latter at least in this, that the lines in the figure are transposed.

(124) There is also another representation of propositions, namely, by means of numbers.* If we put numbers in place of terms, the universal affirmative, or 'A is B' means: A (or, at any rate, the square or the cube of A) can be divided by B. Here, indeed. A, A^2, A^3 are considered as the same.*

(125) The particular affirmative 'some A is B', means that A multiplied by B, i.e. AB, can be divided by B. It is understood, of course, that AB can always be divided by [B], unless [B] is eliminated in AB, if, for instance, A means $\frac{C}{B}$ and C cannot be divided by B.*

(126) The particular negative is that it is false that A can be divided by B, even though perhaps AB can be divided by B.[dd]

(127) The Universal negative is that it is false that AB can be divided by B, the only cause of this being that A contains $\frac{1}{B}$. Thus, properly speaking, the universal negative is that A contains non-B;[ee] therefore, we infer, by means of a consequence, that the Universal negative is the opposite of the particular affirmative, for if A is divided by B, it cannot happen that A is multiplied by B.*

[dd] Everything can be demonstrated by numbers, if only marks of some sort are employed.
[ee] NB.

negatur, itaque respectu formulae distinguitur divisio seu ablatio a negatione, a parte rei non distinguetur.

A = A	vera	A = A : A	falsa
A = A		A non = A : A	
A = AB	univ. Aff.	vel A : B non = A : B	
		seu A : B est falsum	
A = A : B	univ. Neg.	vel AB non = AB	
		seu AB est falsum	
AB = AB	part. Aff.	vel A non = A : B	
A : B = A : B	part. Neg.	vel A non = AB.	

Intelligo hic quendam hominem esse doctum si modo id possibile sit, hoc enim loco nos notiones abstractas, non experimenta consideramus. Si enim possibile sit A = BY, utique istud BY est quoddam B quod est A. Itaque si particularis affirmativa est falsa, impossibile est dari talem notionem.

Videtur optimum, ut prius definiamus particulares, nempe AB est notio vera seu AB = AB est part. Aff.

Et A : B est notio vera seu A : B = A : B est partic. Neg.
Cum vero dicimus AB esse falsam notionem, seu negamus part. Aff., fit univ. Neg.
Cum dicimus A : Besse falsam notionem seu A : B non = A : B, fit Univ. Aff.
Hinc statim patet conversio simpliciter univ. Neg. et part. Aff. Sed ex his demonstrandum jam esse A = AB si A : B non = A : B, et esse A = A : B si AB non = AB.[58])

(128) Habemus ergo has expressiones A = AB est universalis affirmativa.[59] AB = AB est particularis affirmativa; nam et hoc falsum est, si particularis affirmativa sit falsa, quia tunc AB est terminus impossibilis,

---

[58] NB.
[59] Senarius est senarius, omnis homo est animal et sic designari posset AY = AZ, sed hoc puto inutile.

⟨We must distinguish negation from division. With division we have the omission of some term, but it is not for this reason negation, except, indeed, in the case of infinite terms, since it denies what does not inhere in them. Thus, as regards a formula, division or taking-away is distinguished from negation, but from the point of view of the thing itself it will be not distinguished.

A = A	true	A = A : A	false
A = A		A non = A : A	
A = AB	univ. affirmative	or A : B non = A : B	
		or A : B is false	
A = A : B	univ. negative	or AB non = AB	
		or AB is false	
AB = AB	part. affirmative	or A non = A : B	
A : B = A : B	part. negative	or A non = AB.	

I mean here that some man is learned, if only this were possible, for we are considering here abstract concepts, not experiences. For if A = BY is possible, then obviously this 'BY' is some B which is A. Therefore, if the particular affirmative is false, it is impossible that there should be such a notion.

It seems best that we should define the particular propositions first, namely 'AB is a true notion', i.e. 'AB = AB is a particular affirmative'.

And 'A : B is a true notion', i.e. 'A : B = A : B is a particular negative'.

Surely, when we say that AB is a false notion, i.e. we deny the particular affirmative, a universal negative results.

When we say that AB is a false notion, i.e. A : B non = A : B, the universal affirmative results.

Hence, from this we immediately have the simple conversion of the universal negative and of the particular affirmative. But from this we have now to demonstrate that A = AB if A : B non = A : B and that A = A : B if AB non = AB.[ff]⟩

(128) We have, therefore, these expressions: 'A = AB', is the universal affirmative proposition.[gg] 'AB = AB' is the particular affirmative; this too is false, indeed, if the particular affirmative is false, because then 'AB' is an

---

[ff] NB.
[gg] *A six-syllable line is a six-syllable line, every man is an animal*, and suchlike, can be denoted as AY = AZ, but I consider this useless.

quia A continet non-B. A = A non-B est universalis negativa. Unde sequitur falsam esse particularem affirmativam, seu AB esse impossibilem terminum, vel potius falsum (si enim demonstrari hoc perfecte non possit resolvendo in infinitum falsus est, non impossibilis). Denique particularis negativa est A non-B = A non-B. Quae si falsa sit A non-B est impossibilis seu A continet B. Et hoc didici ex considerando numeros. Atque ita tandem plane eliminavimus indefinitam Y. Idque ex numeris didicimus.[60]

(129) Omnia per numeros demonstrari possunt, hoc uno observato, ut AA et A aequivaleant, et ut $\frac{A}{A}$ non admittatur. Quia multiplicatio hoc loco repraesentat complexum notionum, si autem notio aliqua sibi ipsi directe adjiciatur ut Homo homo, nihil aliud fit quam Homo. Divisio autem repraesentat negationem unius de alio, quando scilicet exacte non procedit. Itaque quando A dividi potest exacte per B, ⟨seu quando A continet B⟩, tunc repraesentatur propositio Universalis affirmativa A est B. Quando A dividi potest exacte per non-B seu per $\frac{1}{B}$ ⟨seu quando A continet fractionem $\frac{1}{B}$ (quae repraesentat non-B)⟩ repraesentatur Universalis negativa. At quando A non dividitur exacte per B, oritur particularis Negativa, et quando A non dividitur exacte per $\frac{1}{B}$ oritur particularis affirmativa. Ita arcanum illud detexi, cui ante aliquot annos frustra incubueram.[61]

(130) Vera autem propositio est quae probari potest. Falsa quae non est vera. Impossibilis quam ingreditur terminus [contradictorius].[62] Possibilis quae non est impossibilis. An igitur omnis universalis negativa impossibilis? Ita esse videtur, quia intelligitur de notionibus non de rebus existentibus ut si dico Nullum hominem esse animal, non id intelligo tantum de existentibus hominibus, ⟨sed hinc sequetur quod de singulari aliquo ut Petro negetur, necessario de eo negari⟩. Igitur negandum est omnem Universalem Negativam esse impossibilem, et ad objectionem responderi potest, A continere non-B, probari vel demonstratione seu resolutione perfecta, vel non nisi resolutione in infinitum continuabili seu semper imperfecta. Itaque certum est quidem non vero necessarium, quia nunquam reduci potest ad identicam vel oppositam ad contradictoriam.

---

[60] NB.   [61] NB.   [62] [L: contradictoribus].

impossible term, since A contains not-B. 'A = A non-B' is a universal negative proposition. From this it follows that the particular affirmative is false, i.e. that AB is an impossible, or rather false, term (for if this cannot be demonstrated perfectly by means of an infinite resolution, it is false, not impossible). Finally, the particular negative is 'A non-B = A non-B'; and if this is false, then A non-B is impossible, i.e. A contains B. I have learned this from considering numbers. And thus, finally, we have wholly eliminated the indefinite Y, and this too we have learnt from numbers.[hh]

(129) Everything can be demonstrated by numbers, if only we observe this, that AA and A are equivalent and $\frac{A}{A}$ is not admitted, since multiplication here represents a complex of notions, but if some notion is directly joined to itself, such as 'man man', nothing obtains beside 'man'. Division, however, represents the negation of one with regard to the other, when it does not proceed exactly. Thus, when A can be divided exactly by B, ⟨i.e. when A contains B,⟩ the universal affirmative proposition 'A is B' is represented. When A can be divided exactly by not-B, i.e. by $\frac{1}{B}$, ⟨i.e. when A contains the fraction $\frac{1}{B}$, which represents not-B,⟩ the universal negative is represented. When A is not divided exactly by B, instead, the particular negative arises; and when A is not divided exactly by $\frac{1}{B}$, the particular affirmative arises.* Thus, I have discovered that secret over which I had brooded in vain formerly for several years.[ii]*

(130) A true proposition is one that can be proved, and false proposition one that is not true. An impossible proposition is one into which a [contradictory] term enters; a possible proposition is one that is not impossible. Is every universal negative proposition then impossible? This seems to be the case because it is understood with regard to notions, not to existing things. Thus, if I say that no man is an animal, I do not mean this of existing men alone, ⟨but it will follow from this that what is denied of some singular such as Peter, is necessarily denied⟩. Therefore, it must be denied that every universal negative is impossible, and to the objection we can reply that that A contains non-B is proved either by a demonstration, i.e. a complete resolution, or only by a resolution continuable to infinity, i.e. always incomplete. Thus, it is certain but not necessary, because it can never be reduced to an identical proposition, or its opposite to a contradictory one.

[hh] NB.   [ii] NB.

(130[bis]) Verum igitur est quod probari potest, seu cujus ratio reddi potest, resolutione. Falsum quod contra. Necessarium est quod resolutione reducitur ad identicum. Impossibile est quod resolutione reducitur ad contradictorium. Falsus est terminus vel propositio qui continet opposita utcunque probata. Impossibilis qui continet opposita per reductionem ad finitos probata. Ita ut A = AB si probatio facta est per resolutionem finitam, distingui debeat ab A = AB, si probatio facta est per resolutionem ad infinitum, inde jam oritur illud de Necessario, possibili, impossibili et contingente.

(131) Dupliciter fit resolutio, vel conceptuum in mente, sine experimento (nisi reflexivo quod ita concipiamus) vel perceptionum seu experientiarum. Prior probatione non indiget, nec praesupponit novam propositionem et hactenus verum est quicquid clare et distincte percipio est verum, posterior praesupponit veritatem experimenti. In Deo sola resolutio propriorum requiritur conceptuum, quae tota fit simul apud ipsum. Unde ille novit etiam contingentium veritates, quarum perfecta demonstratio omnem finitum intellectum transcendit.

(132) Omnis propositio vera probari potest, cum enim praedicatum insit subjecto, ut loquitur Aristoteles, seu notio praedicati in notione subjecti perfecte intellecta, involvatur, utique resolutione terminorum in suos valores, seu eos terminos quos continent, oportet veritatem posse ostendi.

(133) Propositio vera necessaria, probari potest reductione ad identicas, vel oppositae reductione ad contradictorias; unde opposita dicitur impossibilis.

(134) Propositio vera contingens non potest reduci ad identicas, probatur tamen, ostendendo continuata magis magisque resolutione, accedi quidem perpetuo ad identicas, nunquam tamen ad eas perveniri. Unde solius Dei est, qui totum infinitum Mente complectitur nosse certitudinem omnium contingentium veritatum.

(135) Hinc veritatum necessariarum a contingentibus idem discrimen est, quod Linearum occurrentium, et Asymptotarum, vel Numerorum commensurabilium et incommensurabilium.

(130[bis]) The true, therefore, is what can be proved, i.e. that whose reason can be given by means of resolution. In case of the false, the contrary holds. Necessary is what by means of resolution is reduced to an identity. Impossible is what by means of resolution is reduced to a contradiction. A term or a proposition is false if it contains opposites, however they are proved, and it is impossible if it contains opposites that are proved by reduction to a finite number of terms. Thus, for instance, A = AB, if the proof has been made by means of a finite resolution, must be distinguished from A = AB, if the proof has been made by an infinite resolution, from which there already arises what we have said about the Necessary, possible, impossible, and contingent.

(131) A resolution is performed in two different ways: either by means of concepts in our mind, without experience (with the only exception of the reflexive experience that we are conceiving in such and such a way), or by means of perceptions or experiences. The former does not need any proof, nor does it presuppose a new proposition and up to this point it is true that whatever I perceive clearly and distinctly is true; the latter presupposes the truth of an experience. In God, only the resolution of his own concepts is required, the whole of which takes place at once in him. Thus, he knows even the truths of contingent propositions, whose complete demonstration transcends every finite intellect.*

(132) Every true proposition can be proved, for since, as Aristotle says, the predicate is in the subject or the notion of the predicate is involved in the notion of the subject understood completely, it must be possible for the truth to be shown, at any rate by means of a resolution of the terms into their values, i.e. of those terms that they contain.*

(133) A true necessary proposition can be proved by reduction to identical propositions, or by reduction of its opposite to contradictory propositions, whence its opposite is called impossible.

(134) A true contingent proposition cannot be reduced to identical propositions: it is proved, however, by showing that if the resolution is continued further and further, it incessantly approaches the identical propositions, but never reaches them. Therefore, only God, who embraces the whole infinite in his mind, has the power of knowing all contingent truths with certainty.*

(135) Therefore, the difference between necessary and contingent truths is the same as that between intersecting and asymptotic lines or commensurable and incommensurable numbers.

(136) At difficultas obstat: possumus nos demonstrare lineam aliquam alteri perpetuo accedere licet Asymptotam, et duas quantitates inter se aequales esse, etiam in asymptotis, ostendendo progressione utcunque continuata, quid sit futurum. Itaque et homines poterunt assequi certitudinem contingentium veritatum. Sed respondendum est, similitudinem quidem esse, omnimodam convenientiam non esse. Et posse esse respectus, qui utcunque continuata resolutione, nunquam se, quantum ad certitudinem satis est, detegant, et non nisi ab eo perfecte perspiciantur, cujus intellectus est infinitus. Sane ut de asymptotis et incommensurabilibus ita et de contingentibus multa certo perspicere possumus, ex hoc ipso principio, quod veritatem omnem oportet probari posse, unde si omnia utrobique se habeant eodem modo in Hypothesibus, nulla potest esse differentia in conclusionibus, et alia hujusmodi, quae tam in necessariis quam contingentibus vera sunt, sunt enim reflexiva. At ipsam contingentium rationem ⟨plenam reddere⟩ non magis possumus, quam asymptotas perpetuo persequi et numerorum progressiones infinitas percurrere.

(137) Multa ergo arcana deteximus magni momenti ad analysin omnium nostrarum cogitationum, inventionemque et demonstrationem veritatum. Nempe quomodo omnes veritates possint explicari numeris. Quomodo veritates contingentes oriantur, et quod naturam quodammodo habeant numerorum incommensurabilium. Quomodo veritates absolutae et Hypotheticae unas easdemque habeant leges, iisdemque generalibus theorematibus contineantur, ita ut omnes Syllogismi fiant Categorici. Denique quae sit origo Abstractorum, quod postremum nunc paulo distinctius explicare operae pretium erit.

(138) Nempe si propositio A est B consideretur ut terminus, quemadmodum fieri posse explicuimus, oritur abstractum, nempe τὸ $\overline{A \text{ esse } B}$, et si ex propositione A est B sequatur propositio [D est C],[63] tunc inde fit nova propositio talis: τὸ $\overline{A \text{ esse } B}$ est ⟨vel continet⟩ τὸ [$\overline{D \text{ esse } C}$],[64] seu Beitas ipsius A, continet Ceitatem ipsius D, seu Beitas ipsius A est Ceitas ipsius D.[65]

---

[63] [L: C est D].   [64] [L: $\overline{C \text{ esse } D}$].

[65] [As Malink and Vasudevan (2016: 694) remark, Leibniz here has been forced to invert 'C' and 'D' to avoid the expression 'Deitas', 'which could be interpreted as deity rather than D-ness'.]

(136) But we are confronted with a difficulty: we can demonstrate that a line, an asymptote, for instance, incessantly approaches another and that two quantities (also in the case of asymptotes) are equal to each other, by showing what will be the case, once the progression is continued as far as we please. Thus, human beings also will be able to gain certitude of contingent truths. But we must reply that there is indeed a similitude here, but not an agreement in all respects. And that there can be respects which, by a resolution continued as far as we please, will never reveal themselves with a sufficient degree of certainty, and which are seen perfectly only by him whose intellect is infinite. Of course, as with asymptotes and incommensurables, so with contingent things we may observe many things with certainty, from the very principle that every truth must be provable. Consequently, if all things are alike on one side and on the other of our hypotheses, then there can be no difference in the conclusions. Analogously, we may observe other things of this sort, which are true in the case of necessary and of contingent propositions as well, because they are reflexive. But we can no more give the full reason of contingent things than we can forever follow the asymptotes and go along the infinite progressions of numbers.

(137) Therefore, we have disclosed many secrets of great importance for the analysis of all our thoughts, and for the discovery and demonstration of truths. We have discovered how all truths can be explained by numbers; how contingent truths arise and that they have, in a certain sense, the nature of incommensurable numbers; how absolute and hypothetical truths have one and the same laws and are contained in the same general theorems, so that all syllogisms become categorical. Finally, we have discovered what the origin of abstract terms is, and now it will be worth our while to explain this last point a little more clearly.*

(138) For if the proposition 'A is B' is considered as a term, as we have explained that it can be, there arises an abstract term, namely 'A's being B'. And if from the proposition 'A is B' the proposition 'C is D' follows, then from this there comes about a new proposition of the following kind: 'A's being B is or contains C's being D', i.e. 'the B-ness of A contains the D-ness of C', or 'the B-ness of A is the D-ness of C'.*

(139) Generaliter autem si dicatur: aliquid esse B, tunc ipsum hoc: aliquid esse B est nihil aliud quam ipsa Beitas. Sic τὸ aliquid esse animal nihil aliud est quam animalitas. At τὸ Hominem esse animal est Animalitas hominis. Unde habemus originem tam abstracti quam talis obliqui.

(140) At per quale abstractum exprimetur τὸ Omnis Homo est animal? An per hoc: Animalitas omnis hominis? Quae longe utique differt ab omni animalitate hominis. Nam modo aliquis homo sit doctus, omnis doctrina hominis est terminus verus; at nisi omnis homo sit doctus, eruditio omnis hominis est terminus falsus. ⟨Nisi quis intelligat terminum exclusive, ut aliquando Geometrae, quando sub omni moto id cujus celeritas est infinite parva, seu quod quiescit.⟩ Videtur eruditio omnis hominis, etiam efferri posse eruditio humanitatis. Sed hoc tamen nolim, si insistimus supra dictis, quod humanitas alicujus nihil aliud sit quam τὸ aliquid esse [hominem] .⁶⁶

(140[⟨bis⟩]) An quia ex eo quod quidam homo est doctus, sequitur: quoddam doctum est homo: dicere licebit: doctrina hominis est humanitas docti? Ita puto.

⟨(141) Quomodo explicabimus quantitatem in abstractis, verbi gratia quando A est duplo calidius ipso B, seu quando calor ipsius A est duplus caloris ipsius B? Scilicet τὸ A esse calidum est calor ipsius A. Itaque si τὸ A esse calidum, sit ad τὸ B esse calidum, ut 2 ad 1, erit calor ipsius A duplus caloris ipsius B. Sed porro videndum est, quomodo τὸ A esse calidum possit [esse] ad τὸ B esse calidum ut numerus ad numerum. Hoc ergo contingit cum causa quae A esse calidum uniformi actione efficit, tali actione adhuc semel continuata efficiat B esse calidum, vel si signum ex quo cognoscimus aliquid esse calidum sit continuum, et in uno alterius duplum. Sed in his multa opus est circumspectione, unde thermometra etsi signa sint graduum caloris, non tamen sunt aequaliter dividenda.⟩

(142) Sed quomodo abstractis efferemus propositiones negativas; ut quidam Homo non est doctus? nempe ut negatio hominis est non-humanitas ita negatio doctrinae hominis est non-doctrina

---

⁶⁶ [L: animal].

(139) In general, however, if we say that something is B, then this 'something being B' is nothing else than the B-ness itself. Thus, 'something being animal' is nothing else than 'animality'. Whereas 'man's being animal' is the animality of man. From this originate both the abstract term and such an oblique.

(140) But by means of what abstract will '[every man is an animal]' be expressed? Maybe by means of 'animality of every man'? Which is surely very different from every animality of man. If only some man is learned, indeed, 'every learning of man' is a true term; but unless every man is learned, 'the learning of every man' is a false term— ⟨that is, unless someone understands the term exclusively, as sometime geometers are doing, when under everything in motion they include that whose speed is infinitely small, or which is at rest⟩. It seems that 'the learning of every man' can be transformed into 'the learning of humanity'. But I would prefer not to do this, if we rely on what has been said before, because 'the humanity of something' is nothing else than 'something being a man.'

(140 [⟨bis]⟩) Or, because from the fact that some man is learned it follows that 'something learned is a man', are we authorized to say 'the learning of a man is the humanity of something learned'? I think so.

⟨(141) How shall we explain quantity in abstract terms, for instance, when A is twice as hot as B, i.e. when the heat of A is twice the heat of B? Obviously, 'A's being hot' is the heat of A; thus, if 'A's being hot' is to 'B's being hot' as 2 is to 1, the heat of A will be twice that of B. But we must see further how 'A's being hot' can be to 'B's being hot' as one number to another. This so happens because the cause which makes A be hot with a uniform action, when such an action is once continued, makes B be hot; or, if the sign by which we know that something is hot is continuous and in the one case is twice what it is in the other. But we need to be very cautious here and even though thermometers are signs of degrees of heat, they are not to be divided equally.⟩

(142) But how may we express negative propositions in abstract terms, for instance, 'some man is not learned'? Certainly, as the negation of 'man' is 'non-humanity', so the negation of the learning of man is the

hominis. Et si dicatur nullus homo est lapis; abstractum ejus seu τὸ nullus homo *est* [lapis],[67] efferendum erit, non-lapideitas omnis hominis; an vero dicere licebit: lapideitas nullius hominis? seu lapideitas non-hominis? Non puto; neque enim id exprimit nullum hominem esse lapidem.

(143) Illud jam videndum est, an cum abstractorum praedicationibus consentiat haec doctrina, et quidem viriditas est color, praedicatio bona est, cur ita? An quia sequitur qui est viridis, eundem esse coloratum? Sed videamus an non exempla sint in contrarium: Circulus est uniformis, item circulus est planum. Non tamen dici potest uniformitatem esse planitiem, quia ex uniformitate non sequitur planities. An vero dicemus uniformitas circuli est planities? Sane videtur ex propositione Circulus est uniformis sequi Circulus est planum. Equidem verum est non sequi ex hac propositione magis quam ex quavis alia de circulo. An ergo videntur praedicationes abstractorum non tantum consequentiam postulare, sed et aliquid praeterea. Quid ergo quia Omnis circulus est uniformis, seu quia si A est circulus, sequitur quod A est uniformis, licebitne ideo dicere Circularitas est uniformitas? Ergo pari jure dicere licebit: Circularitas est planities. Et proinde dici poterit: Quoddam quod est uniformitas est planities. In quibus tamen haereo adhuc nonnihil. Sane si idem sit, uniformitas, quod τὸ uniforme esse, et planities quod το planum esse, an verum est aliquando quod τὸ A uniforme esse, sit τὸ A planum esse. Ita puto, quando ipsum A est uniforme respectu unius centri. Unde dici poterit Uniformitas respectu unius centri est planities seu existentia in plano. Et vero quemadmodum in concretis sunt praedicationes per accidens, cum Musicus est poeta, non video cur non et admittantur in abstractis, ut uniformitas aliqua sit planities. Recte igitur dicemus uniformitatem circuli esse planitiem, et proinde poterimus insistere regulae generali. Sed quomodo jungemus haec in circularitate. An quia dicimus circularitas est uniformitas, et circularitas est planities dicere licebit uniformitas est circularitas planities? Et an non videntur confundi officia praedicamentorum, ut dici possit quaedam qualitas est quantitas, cum aliquando ex eo quod quis est qualis sequitur eum esse

---

⁶⁷ [L: doctus].

non-learning of man. And if someone says 'no man is a stone', the abstract term which expresses it, i.e. 'no man is a stone' will be expressed as the 'non-stoneness of every man'; or will it be permissible to say 'stoneness of no man'? or 'stoneness of non-man'? I do not think so; this, indeed, does not express that no man is a stone.

(143) We ought to see now whether this doctrine agrees with the predication of abstract terms. Clearly, 'greenness is a colour' is a good predication, but why is this so? Maybe because it follows that what is green is coloured? But let us see whether or not there are examples to the contrary: a circle is uniform; likewise a circle is plane; yet we cannot say that uniformity is planeness, because from uniformity does not follow planeness. Should we say, then, 'the uniformity of a circle is planeness'? Certainly, it seems that from the proposition 'A circle is uniform', 'A circle is plane' follows. But it is true that it does not follow from this proposition in particular, rather than from any other proposition about the circle. Does it seem, therefore, that predications of abstract terms postulate not only a consequence, but something else as well? What, then? Because every circle is uniform, i.e. because if A is a circle, it follows that A is uniform, are we authorized to say that circularity is uniformity? Then we may say with equal right 'circularity is planeness' and, therefore, we may say even 'some thing which is uniformity is planeness'. About these matters, however, I am still quite perplexed. If 'uniformity' is the same as 'being uniform' and 'planeness' the same as 'being plane', is it sometime true that A's being uniform is A's being plane? I think so, when A is uniform in respect of only one centre. Therefore, it will be possible to say 'uniformity as regards one centre is planeness', i.e. existence in a plane. And just as in the case of concrete terms there are predications *per accidens*, when a musician is a poet, I do not see why they should not be admitted in the case of abstract terms too, so that some uniformity is planeness. Therefore, we shall say correctly that the uniformity of a circle is planeness, and so we shall be able to follow the general rule: but how can we join these properties in circularity? Maybe, because we say 'circularity is uniformity' and 'circularity is planeness', are we authorized to say 'uniformity is circularity planeness'? But in such a case, do not the roles of the categories seem to be confused, given that we can say 'some quality is a quantity', because sometimes from the fact

quantum. Quid hinc? Modo non possit dici omnis qualitas est quantitas.

Videndum an in casu talis propositionis universalis in abstractis sequatur necessitas in concretis, puto ne hoc [⟨quidem⟩] sequi, sunt enim contingentes connexiones semper verae, quae pendent a liberis actionibus.

(144) Propositiones sunt vel Essentiales vel existentiales; et ambae vel secundi vel tertii adjecti. *Propositio essentialis tertii adjecti* ut: Circulus est figura plana. *Propositio essentialis secundi adjecti* ut: figura plana ad unum aliquod punctum eodem modo se habens, est; est, inquam, hoc est intelligi potest, concipi potest, inter varias figuras est aliqua quae hanc quoque naturam habet; perinde ac si diceremus: figura plana ad unum aliquod punctum eodem modo se habens, est ens sive res. Propositio *existentialis tertii adjecti*. Omnis homo est seu existit peccato obnoxius, haec scilicet est propositio existentialis seu contingens. *Propositio existentialis secundi adjecti*: Homo peccato obnoxius est seu existit, seu est ens actu.

(145) Ex omni propositione tertii adjecti fieri potest propositio [secundi][68] adjecti, si praedicatum cum subjecto componatur in unum terminum, isque dicatur esse vel existere, hoc est dicatur esse res sive utcunque, sive actu existens.

(146) Propositio particularis affirmativa, Quoddam A est B transformata in propositionem secundi sic stabit: AB est, hoc est, AB est res nempe vel possibilis vel actualis, prout propositio est essentialis vel existentialis.

(147) Propositio Universalis Affirmativa in propositionem secundi adjecti hoc quidem modo non aeque commode transformatur, nam ex Omne A est B non licet commode facere: Omne AB, est. Cum enim AB sit idem quod BA, pari jure dicere liceret Omne BA, est; et proinde etiam Omne B est A. Itaque sic dicendum erit Omne A continens B, est. Quomodo autem alia ratione propositio universalis affirmativa ad secundi adjecti enuntiationem reducatur mox patebit.

(148) Propositio particularis Negativa Quoddam A, non est B, sic transformabitur in propositionem secundi adjecti: A, non-B; est, hoc est

---

[68] [L: tertii].

that someone is such and such it follows that he is of a certain quantity? And so? Provided that it cannot be said that every quality is a quantity. We must see whether in the case of such a universal proposition in abstract terms necessity follows in the concrete terms. I think that this does not certainly follow, for there are contingent connections which are always true, but which depend upon free actions.

(144) Propositions are either essential or existential, and both are either *secundi adiecti* or *tertii adiecti*. An *essential tertii adiecti* proposition, is, for example, 'A circle is a plane figure'. An *essential secundi adiecti* proposition is, for example, 'a plane figure maintaining the same relation to some one point in the plane is'. I say 'is': that is, it can be understood, it can be conceived, that among various figures there is one which also has this nature, just as we were to say 'a plane figure maintaining the same relation to some one point in the plane is a being or thing'. An *existential tertii adiecti* proposition is 'Every man is who is liable to sin is, i.e. exists', which obviously is an existential or contingent proposition. An *existential secundi adiecti* proposition is 'A man liable to sin is, or exists', i.e. is actually a being.*

(145) From every *tertii adiecti* proposition a [*secundi*] *adiecti* proposition can be made, if the predicate is compounded with the subject into one term and this is said to be or to exist, i.e. is said to be a thing, either in any way whatsoever, or actually existing.

(146) The particular affirmative proposition 'Some A is B', once transformed into a *secundi adiecti* proposition will become 'AB is', i.e. AB is a thing, either possible or actual, depending on whether the proposition is essential or existential.*

(147) The universal affirmative proposition is not so easily transformed, at least by this method, into a *secundi adiecti* proposition. For from 'Every A is B', indeed, it is not easily permitted to make 'Every AB is': AB being the same as BA, we may say with equal right 'Every BA' is, and thus 'Every B is A', as well. Therefore, we must say 'Every A containing B is'. However, it will become evident soon how, on the basis of a different method, a universal affirmative proposition is reduced to a *secundi adiecti* proposition.

(148) The particular negative proposition 'Some A is not B' will be transformed into a *secundi adiecti* proposition in this way: 'A non-B is',

A quod non est B est res quaedam; possibilis vel actualis, prout proposi-tio est essentialis vel [existentialis].[69]

(149) Universalis negativa transformatur in propositionem secundi adjecti per negationem particularis affirmativae. Verbi gratia Nullum A est B, hoc est AB non est, seu AB non est res. Posses etiam sic enuntiare: Nullum A est B, id est: Omne A continens non-B est.

(150) Universalis affirmativa transformatur in propositionem secundi adjecti per negationem particularis negativae, ita ut Omne A est B, idem sit quod: A non-B non est seu non est res, vel etiam (ut dixi n. 147) A continens B est res. Quod tamen posterius ut jam dixi minus aptum est, etsi verum sit, quia est superfluum, jam enim B in A continetur, sed si non omne A sit B, ex AB fit nova res.

(151) Habemus ergo propositiones tertii adjecti sic reductas ad pro-positiones secundi adjecti:

Quoddam A est B dat: *AB est res.*
Quoddam A non est B dat: *A non-B est res.*
Omne A est B dat: *A non-B non est res.*
Nullum A est B dat: *AB non est res.*

(152) Et cum ipsis identicis propositionibus tantum fidi possit in notionibus realibus, adeo ut veritas nulla sine metu oppositi asseri possit nisi de ipsarum notionum realitate saltem essentiali, licet non existen-tiali, constet; ideo licebit propositionum Categoricarum Species quatuor etiam sic exprimere:

Part. Aff.	AB = AB (seu AB et AB coincidunt, hoc est AB est res).
Part. Neg.	A non-B = A non-B (seu A non-B est res).
Univ. Aff.	A non-B non = A non-B (seu A non-B non est res).
Univ. Neg.	AB non = AB (seu AB non est res).

(153) Hoc autem praesupponit negari omnem propositionem, quam ingreditur Terminus qui non est res. Ut scilicet maneat omnem proposi-tionem vel veram vel falsam esse; falsam autem omnem esse cui deest

---

[69] [L: essentialis].

that is: A, which is not B, is some thing, possible or actual, depending on whether the proposition is essential or [existential].

(149) The universal negative is transformed into a *secondi adiecti* proposition by the negation of the particular affirmative. For instance: 'No A is B', i.e. 'AB is not' or 'AB is not a thing'. It could also be stated this way: 'No A is B', i.e. 'Every A containing non-B is'.

(150) The universal affirmative is transformed into a *secundi adiecti* proposition by the negation of the particular negative; thus, for instance, 'Every A is B' is the same as: 'A non-B is not' or 'is not a thing', or even (as I have said in n. 147) 'A containing B is a thing'. The latter, however, as I have already said, even though is true, is less apt, because it is superfluous: B, indeed, is already contained in A, but if not every A is B, a new thing is made with AB.

(151) Therefore, we have *tertii adiecti* propositions reduced in this way to *secundi adiecti* propositions:

'Some A is B'	gives	*AB is a thing*
'Some A is not B'	gives	*A non-B is a thing*
'Every A is B'	gives	*A non-B is not a thing*
'No A is B'	gives	*AB is not a thing.*

(152) And since we can rely on identical propositions only to the extent that we have to do with real notions, so that no truth can be asserted without fear of the opposite, with the exception of those truths concerning at least the essential (not the existential) reality of the notions themselves, it will be permissible for this reason to express as follows the four species of Categorical propositions:

Particular Affirmative	AB = AB (i.e. AB and AB coincide that is : 'AB is a thing').
Particular Negative	A non-B = A non-B (i.e. 'A non-B is a thing').
Universal Affirmative	A non-B not = A non-B (i.e. 'A non-B is not a thing').
Universal Negative	AB not = AB (i.e. 'AB is not a thing').

(153) This, however, presupposes that every proposition in which there enters a term which is not a thing, is denied; so it remains the case that every proposition is either true or false, whereas every one is false which

Constantia Subjecti, seu terminus realis. Hoc tamen nonnihil ab usu loquendi remotum est in propositionibus existentialibus. Sed hoc ego non est cur curem, quia propria signa quaero, non recepta nomina his applicare constituo.

(154) Quod si quis malit signa sic adhiberi, ut AB sit = AB, sive AB sit res sive non, et ut eo casu quo AB non est res possint coincidere B et non-B, scilicet per impossibile, non equidem repugno. Et ita distinguendum erit inter Terminum et Rem seu Ens.

(155) Omnibus ergo expensis fortasse melius erit, ut dicamus semper in characteribus quidem poni posse A = A, licet quando A non est res, nihil inde utiliter concludatur. Itaque si AB sit res poterit inde fieri YA = ZB. Nam AB = R, et AB = RB. Sit B = Y et R = Z, fiet YA = ZB. Et contra YA = ZB. Ergo YAB = ZB. Jam A = R et B = (R) (seu A et B sunt res). Ergo YAB = Z(R). Ergo AB = ((R)).

(156) A = A.

A non = non-A.

AA = A.

(157) A = B est universalis affirmativa reciproca, quae est simplicissima. Coincidit cum non-A = non-B, et si negetur dici poterit A non = B.

(158) D = ZC est Univ. Aff.

(159) YA = ZC est Partic. Aff.

(160) D = non-E Universalis negativa.

(161) XE = non-F particularis negativa.

(162) Supersunt termini quos ingrediuntur non [$\overline{\text{YA}}$],[70] hoc est non tale A (seu quoddam A non) qui differunt a non quoddam. Nempe aliud est dicere falsum esse, quoddam A esse B. Aliud est dicere falsum esse tale A esse B. Unde cum hic oriatur aequivocatio aliqua, satius erit literas Y prorsus eliminare, et hinc orientur tales propositiones.

(163) A = B idem non-A = non-B simplicissimae.

(164) A = AB *universalis Affirmativa.*

(165) AB = AB posito AB esse rem, *particularis Affirmativa,* seu YA = ZB.

---

[70] [L: YA]

lacks an existent subject, i.e. a real term.* This, however, is to some extent remote from the way we usually speak about existential propositions. But this is no reason for concern, because I am seeking appropriate signs, and I do not intend to apply usually accepted names to them.

(154) But if someone prefers to employ signs so that AB is = AB, whether AB is a thing or not, and so that, in the case in which AB is not a thing, B and not-B can coincide, namely through the impossible, obviously I do not object. But in this case we must distinguish between a Term and a Thing or Being.

(155) Therefore, everything considered, it will perhaps be better to say that when we employ symbols, it is always possible to state clearly A = A, even though, when A is not a thing, nothing can be usefully concluded from this. Thus, if AB is a thing, it will be possible to have from this YA = ZB. For AB = R and AB = RB; assuming B = Y and R = Z, we will have YA = ZB. And vice versa: YA = ZB, therefore YAB = ZB. Now, A = R and B = (R) (i.e. A and B are things). Therefore YAB = Z(R). Therefore AB = ((R)).*

(156) A = A.

A not = non-A.

AA = A.

(157) A = B is a reciprocal universal affirmative proposition, and it is the simplest. It coincides with non-A = non-B, and if it is denied, it can be said that A not = B.

(158) D = ZC is a Universal Affirmative proposition.

(159) YA = ZC is a Particular Affirmative proposition.

(160) D = non-E Universal Negative proposition.

(161) XE = non-F is a Particular Negative proposition.

(162) There remain those terms into which there enter 'non [ßY͞A]', i.e. 'not such an A' (i.e. 'some A not'), which are different from 'not some'. Clearly, it is one thing to say that it is false that some A is B, and another to say that it is false that such an A is B. Thus, because some equivocation arises here, it will be preferable to eliminate completely the Y letters, and then the following propositions will arise.*

(163) A = B and non-A = non-B alike, the most simple propositions.

(164) A = AB is the *universal affirmative.*

(165) AB = AB assuming that AB is a thing, is the *particular affirmative,* or YA = ZB.

(166) A = [A] non-B universalis [negativa].[71]

(167) A non-B = A non-B, posito A non-B esse rem, particularis negativa.

(168) Si A non = B tunc vel A non-B erit res, vel B non-A erit res.

(169) AB est res aequivalet Quoddam A est B, et Quoddam B est A.

A non-B est res aequivalet Quoddam A non est B vel Quoddam A est non-B.

A non-B est non res aequivalet Universali Affirmativae: Omne A est B.

AB est non res aequivalet Universali Negativae: Nullum A est B, vel Nullum B est A.

(170) Interim opus est tamen, ut propositionem Quoddam A est B discernamus a propositione: Quoddam B est A, et similiter Nullum A est B, a propositione nullum B est A.

(171) Principia sunt:

Primo A = A.

Secundo non-A = non-A.

Tertio AA = A.

⟨*Quarto* non-non = omissioni ipsius non, ut non-non-A = A.⟩

Quinto Si A = B erit AC = BC.

Sexto Si A = B erit non-A = non-B.

Septimo Si A = B, non erit A = non-B.

Octavo A non-A non est res.

(172) Si A = B erit AB = B. Nam A = B ex hypothesi, ergo AB = BB per princip. quintum, id est per princip. 3 AB = B.

(173) Si A = BC erit AB = BC. Nam A = BC ex hyp. Ergo AB = BBC per princip. quintum, id est per princip. 3 AB = BC.

(174) Si non-A = B erit non-B = A. Nam sit non-A = B ex hyp. erit non-non-A = non-B per princip. sext. Jam non-non-A = A per princip. 4. Ergo A = non-B.

(175) Si A = non-B non erit A = B. Nam sit A = non-B ex hyp. non erit A = non-non-B per princip. [7].[72] Ergo per princip. 4 non erit A = B.

(176) Si A = BC erit A = AC. Nam sit A = BC (per hyp.) erit A = ABC = BCBC = BCC = AC.

(177) Si A = YC erit A = AC, ut ante.

---

[71] [L: A = non-B universalis affirmativa].   [72] [L: 6].

(166) A = [A] non-B is the *universal* [*negative*].

(167) A non-B = A non-B, assuming that A non-B is a thing, is the *particular negative*.

(168) If A not = B, then either A non-B will be a thing or B non-A will be a thing.

(169) 'AB is a thing' is equivalent to 'some A is B' and 'some B is A'.

'A non-B is a thing' is equivalent to 'some A is not B' or 'some A is non-B'.

'A non-B is not a thing' is equivalent to the universal affirmative: 'every A is B'.

'AB is not a thing' is equivalent to the universal negative: 'no A is B', or 'no B is A'.

(170) Meanwhile, however, we need to distinguish the proposition 'some A is B' from the proposition 'some B is A', and similarly the proposition 'no A is B', from the proposition 'no B is A'.

(171) The principles are:*

First	A = A.
Second	non-A = non-A.
Third	AA = A.
⟨*Fourth*	non-non = the omission of 'non', as non-non-A = A.⟩
Fifth	If A = B, there will be AC = BC.
Sixth	If A = B, there will be non-A = non-B.
Seventh	If A = B, there will not be A = non-B.
Eighth	A non-A is not a thing.

(172) If A = B, then there will be AB = B. For A = B *ex hypothesi*, therefore AB = BB by the fifth principle, that is, by principle 3, AB = B.

(173) If A = BC, there will be AB = BC. For A = BC *ex hypothesi*; therefore AB = BBC by the fifth principle, that is, by principle 3, AB = BC.

(174) If non-A = B, there will be non-B = A. For, assume non-A = B *ex hypothesi*: then there will be non-non-A = non-B, by the sixth principle. Now: non-non-A = A, by the fourth principle. Therefore A = non-B.

(175) If A = non-B, there will not be A = B. For, assume A = non-B, *ex hypothesi*; then there will not be A = non-non-B, by the seventh principle. Therefore, by the fourth principle, there will not be A = B.

(176) If A = BC, there will be A = AC. For, assume A = BC, *ex hypothesi*; then there will be A = ABC = BCBC = BCC = AC.

(178) Si A = YC erit ZA = VC. Nam A = YC ex hyp. Ergo ZA = ZYC, sit ZY = V, fiet ZA = VC.

(179) Si A = YC erit VC = ZA, patet ex praecedenti.

(180) Si A = [non $\overline{AC}$][73] erit A = non-C. (Scilicet si A est res.). Hoc accurate demonstrandum.[74]

(181) [non $\overline{AC}$] = Y non-C (= Z non-A).

(182) Si Y non-C = Z non-A, erit = [non $\overline{AC}$]. } haec demonstranda.

(183) non-A non-C = [Y non $\overline{AC}$].

(184) Omnis propositio in sermone usitata huc redit ut dicatur quis terminus quem contineat, et quidem inspicitur [terminus continens],[75] vel absolutus, vel cum addito, et is dicitur continere contentum absolutum.

(185) Non debet in propositionibus proprie occurrere: non omnis, non quidam; haec enim tantum negant propositionem signo omnis aut quidam affectam, non faciunt novum signum non-omnis, vel nonquidam; sic si dicam non, quidam homo est animal idem est quod falsum est, quendam hominem esse animal.

(186) Quidam homo non est lapis significat: quidam homo est nonlapis, istud: Omnis homo non est lapis videtur significare Omnis homo est non-lapis; itaque generaliter sic interpretabimur non ante est quasi praedicatum negativum, sed si τὸ *non* praeponitur signo, intelligemus propositionem negari.

(187) Jam supra monui quae ad propositiones pertinent sic posse illustrari et quasi ad numeros revocari, ut concipiamus Terminum seu Notionem instar fractionis verbi gratia ab non-l non-m = H quod significat H continere a, et b, sed idem H continere non-l et non-m; observando tantum ut aa idem sit quod a, ⟨et non-a non-a idem quod non-a et non-non-a idem quod a⟩, et ut nunquam idem terminus contineat simul a et non-a, seu ut terminus qui continet a non dicatur continere non-a vel contra. ⟨Denique qui continet ab continere etiam a, et qui continet non-a continere etiam [non $\overline{al}$].⟩[76]

---

[73] [L: non AC].   [74] NB.   [75] [L: termini continentis].
[76] [The nr. 188 is deleted: at the end of it Leibniz has written (and not deleted):] Si a non sit = a, terminus a est falsus.

(177) If A = YC, there will be A = AC, as before.

(178) If A = YC, there will be ZA = VC. For A = YC, *ex hypothesi*; therefore ZA = ZYC, and assuming ZY = V, there will be ZA = VC.

(179) If A = YC, there will be VC = ZA. This is evident from the preceding.

(180) If A = [non-$\overline{AC}$], there will be A = non-C (namely, if A is a thing). This must be carefully demonstrated.^{jj}

(181) [non $\overline{AC}$] = Y non-C (= Z non-A). ⎫
(182) If Y non-C = Z non-A,   ⎬ These need to be
there will be = [non $\overline{AC}$].   ⎪ demonstrated.
(183) non-A non-C = [Y non $\overline{AC}$]   ⎭

(184) Every proposition usually employed in speech comes to this, that it is said which term contains which; moreover the [containing term], whether it is absolute or with something added, is examined, and is said to contain an absolute content.

(185) 'Not every' and 'not some' ought not properly to occur in propositions, for these only deny the proposition affected by the sign 'every' or 'some', and do not make a new sign 'not-every', or 'not-some'. Thus, if I say 'not, some man is an animal', it is the same as 'it is false that some man is an animal'.

(186) 'Some man is not a stone' means 'some man is a non-stone', and this: 'Every man is not a stone' seems to mean 'Every man is non-stone'. Thus, in general, we shall interpret 'not' after 'is' as if it were a kind of negative predicate, but if *not* is prefixed to the sign, then we shall understand that the proposition is denied.

(187) I have remarked above that those things that concern propositions can be illustrated by and, as it were, reduced to numbers, so that we conceive a term or notion as a fraction, for example: 'ab non-l non-m = H', which means that H contains 'a' and 'b', but also that the same H contains 'non-l' and 'non-m' (sticking to the only rules according to which 'aa' is the same as 'a', ⟨and 'non-a non-a' is the same as 'non-a'; that 'non-non-a' is the same as 'a'⟩ and that the same term never contains 'a' and 'non-a' at the same time, or that the term which contains 'a' is not said to contain 'non-a', or conversely). ⟨Finally, the term which contains 'ab' contains also 'a', and the term which contains 'not-a' also contains '[not $\overline{a}$]'.⟩^{kk}

---

^{jj} NB.
^{kk} In the margin, near the deleted Nr. 188: 'If *a* is not = *a*, the term *a* is false.'

(189) Principia ergo haec erunt:
Primo aa = a (unde patet etiam non-b[77] = non-b, si ponamus non-b = a).
Secundo non-non-a = a.

Tertio non idem terminus continet a et non-a seu si unum est verum alterum est falsum, ⟨aut certe terminus ipse talis dicetur non verus sed falsus⟩.

Quarto A continere l idem est quod A esse = xl.

Quinto non-a continet non $\overline{ab}$, seu si l continet a, non-a ⟨continebit non-l⟩.

Sexto Quaecunque dicuntur de Termino continente terminum, etiam dici possunt ⟨de propositione ex qua sequitur alia propositio⟩.

Septimo Quicquid ex his principiis demonstrari non potest, id non sequitur vi formae.

(190) *Universalis Affirmativa* Omne A est L idem est quod A continere L seu A = XL.

Particularis Affirmativa: quoddam A est L, idem est quod A cum aliquo addito sumtum continere L, verbi gratia AB continere L posito B = LX, vel AN continere L, posito esse L = MN, et A = BM, nam ita fiet AN = BMN = BL. Proinde etiam quoddam A est L idem est quod AL continet L, seu AL = AL; posito scilicet AL esse rem seu terminum verum qui non implicat opposita ut X non-X.

Universalis Negativa Omne A est non-B seu A continet non-B seu A = X non-B.

Particularis Negativa Quoddam A est non L seu AX continet non-L seu AX = Z non-L seu et A non-L continet non-L, seu A non-L = A non-L, posito A non-L esse terminum verum qui non implicat opposita.

(191) Si vera est Universalis Affirmativa, vera etiam est Particularis Affirmativa, seu si A continet B, etiam quoddam A continet B. Nam A = XB per princip. 4. Ergo ZA = ZXB (ex natura coincidentium). Sit ZX = V (ex arbitrio) fiet ZA = VB.

---

[77] [As Malink and Vasudevan (2016: 698) remark, the correction proposed by the Academy edition and some commentators (Schupp 1982: 122–3; Rauzy 1998: 294–5; Mugnai 2008: 110): 'non-b [non-b] = non-b' is misplaced. As they argue:

Leibniz's intent in §189.1 is to derive the reflexivity of coincidence from the principle AA = A. In his initial list of principles in §171, the first three principles stated are: A = A, non-A = non-A, and AA = A. In his revised list of principles in §189, Leibniz omits the first two of these principles. This strongly suggests that he regarded reflexivity as derivable from other principles, which is just what the unaltered text of §189.1 implies.]

(189) The principles, therefore, will be these:

First, aa = a (from which it is also clear that non-b = non-b, if we assume that non-b = a).

Second, non-non-a = a.

Third, the same term does not contain a and non-a, i.e. if the one is true, the other is false, ⟨or surely this very term will itself not be called true, but false.⟩*

Fourth, that A contains l is the same as A is = xl.

Fifth, not-a contains not a̅b̅; i.e. if l contains a, non-a ⟨will contain non-l.⟩

Sixth, Whatever is said of a term containing a term can also be said ⟨of a proposition from which another proposition follows⟩.

Seventh, Whatever cannot be demonstrated from these principles does not follow in virtue of the form.

(190) The *universal affirmative* 'every A is L' is the same as that A contains L or A = XL.

The *particular affirmative*: 'some A is L' is the same as that A taken with something added contains L; for example, that AB contains L, if we assume that B = LX, or that AN contains L, if we assume that L = MN, and A = BM, for thus we will have AN = BMN = BL. Consequently, even 'some A is L' is the same as 'AL contains L', or AL = AL, assuming, obviously, that AL is a thing, i.e. a true term which does not imply opposites, as X non-X.

The *universal negative*: 'every A is non-B', i.e. A contains non-B, i.e. A = X non-B.

The *particular negative*: 'Some A is not-L', i.e. AX contains non-L, i.e. AX = Z non-L or even A non-L contains non-L, or A non-L = A non-L, if we assume that non-L is a true term which does not imply opposites.*

(191) If the universal affirmative is true, the particular affirmative is also true, i.e. if A contains B, then some A contains B as well. For A = XB by the fourth principle; therefore ZA = ZXB (from the nature of coincidents). If we assume (arbitrarily) ZX = V, then we will have ZA = VB.

(192) In terminis veris propositio Universalis Affirmativa et particularis negativa non possunt esse simul verae, sit enim A = XL et VA = Z non-L, fiet AVA seu VA = AZ non-L = XLZ non-L, qui terminus est falsus.

(193) Eaedem non possunt simul esse falsae. Sit A non = AL, et A non-L non = A non-L, erit A non-L terminus falsus, ergo A = AL.

(194) Terminus falsus est ⟨qui continet oppositos⟩ A non-A. Terminus verus est non-falsus.

(195) Propositio est quae pronuntiat quis terminus in alio contineatur aut non contineatur. Unde etiam propositio affirmare potest terminum aliquem esse falsum, si dicat in eo contineri Y non-Y; et verum si neget. Propositio etiam est quae dicit utrum aliquid alteri coincidat aut non coincidat, nam quae coincidunt in se invicem continentur.

(196) Propositio falsa est, quae continet oppositas, ut ⊙ et non ⊙.

(197) Ipsa propositio concipi potest instar termini, sic quoddam A esse B, seu AB esse terminum verum, est terminus, nempe AB verum. Sic Omne A esse B, seu A non-B esse falsum, seu A non-B falsum est terminus novus. Sic Nullum A esse B seu AB esse falsum est terminus novus.

(198) Principia:

1°. ⟨Coincidentia sibi substitui possunt⟩.

2do. AA = A.

3°. non-non-A = A.

4°. Falsus seu non verus est terminus qui continet A non-A; verus qui non continet.

5°· Propositio est quae termino addit quod sit verus vel falsus, ut: si A sit terminus eique ascribatur A verum esse, A non verum esse, solet etiam simpliciter dici A esse, A non esse.

6°. Veri seu τοῦ esse adjectio relinquit, at falsi seu τοῦ non esse in oppositum mutat; itaque si verum aut falsum quid esse verum dicatur, manet verum aut falsum; sin verum aut falsum esse falsum dicatur, fit ex vero falsum, ex falso verum.

7°. Propositio ipsa fit Terminus si termino ipsi adjiciatur verum aut falsum; ut sit A terminus, et A *est* vel *A verum est*, sit propositio, *A verum*, seu *A verum esse*, seu *A esse* erit terminus novus, de quo rursus fieri potest propositio.

(192) In the case of true terms, a universal affirmative and a particular negative proposition cannot be true at the same time, since, suppose A = XL and VA = Z non-L: then we shall have AVA or VA = AZ not-L = XLZ not-L, which is a false term.

(193) These same propositions cannot be false at the same time. For, suppose A not = AL, and A non-L not = A non-L, then A non-L will be a false term; therefore, A = AL.

(194) A false term is one ⟨which contains opposite terms⟩, A non-A. A true term is non-false.

(195) A proposition is that which states what term is or is not contained in another. Thus, a proposition can also affirm that some term is false, if it says that Y non-Y is contained in the term; and true if it denies this. A proposition is also that which says whether or not one thing coincides with another, for those things that coincide are contained in each other reciprocally.

(196) A proposition is false which contains opposites, as ⊙ and non ⊙.

(197) A proposition itself can be conceived as a term; thus, that some A is B, or that AB is a true term, is a term, namely AB true. Hence, that every A is B, or that A not-B is false, or 'A non-B false' is a new term. Thus, that no A is B or that AB is false is a new term.

(198) Principles:

1st. ⟨Coincidents can be substituted for one another⟩.

2nd. AA = A.

3rd. non-non-A = A.

4th. False, i.e. not true, is that term which contains A non-A; true is that term which does not contain it.

5th. A proposition is what adds to a term that it is true or false; thus, for example, if A is a term and to it is ascribed A's being true or A's being not true, we simply tend to say that A is or A is not.

6th. The addition of 'true' or 'to be' leaves things as they are, but that of 'false' or 'not to be' changes them into their opposite. Thus, if we say that something which is true or false is true, it remains true or false; but if we say that something true or false is false, then false is made from true and true from false.

7th. A proposition itself becomes a term if true or false is added to it. Suppose, for example, that A is a term and that the proposition is 'A is' or 'A is true'; 'A true', or 'A's being true' or 'A's being' will be a new term, from which again a proposition can be made.

8°. Propositionem ex propositione sequi nihil aliud est quam consequens in antecedenti contineri ut terminum in termino, atque hac methodo reducimus consequentias ad propositiones, et propositiones ad terminos.

9°. A continere l idem est quod A = xl.

(199) Propositio, *particularis affirmativa*: AB est.

Particularis negativa A non-B, est.

Et posito A et B esse *Universalis affirmativa* A non-B non est.

Universalis Negativa: AB non est.

Hinc statim patet nec numero plures dari, et quaenam earum sint oppositiones et conversiones. Nam P. A. et U. N. opponuntur, item P. N. et U. A. Patet etiam in propositione AB est vel AB non est utrumque terminum eodem modo se habere, et ideo locum habere conversionem simpliciter. Addi posset: non-A non-B est, vel non-A non-B non est; sed nihil differt a LM est, vel LM non est, posito non-A esse L et non-B esse M. U. A. seu A non-B non est, idem est quod A continet B, nam A non continere B est idem quod A non-B esse verum. Ergo A continere B idem quod A non-B esse non verum.[78]

⟨(200) Si dicam AB non est, idem est ac si dicam A continet non-B, vel B continet non-A, seu A et B sunt inconsistentia. Similiter si dicam A non-B non est, idem est ac si dicam A continet non-non-B seu A continet B, et similiter non-B continet non-A.

His ergo paucis formae fundamenta continentur.⟩

---

[78] Omne B est C. B non-C non est.
Omne A est B. A non-B non est.
Omne A est C. A non-C non est.
Sed haec consequentia ex meris negativis etsi proba sit, non tamen apparet, nisi reducta ad affirmativas. Unde apparet hanc reductionem universalium ad negativas non esse adeo naturalem. Quemadmodum A continet B et B continet C etiam A continet C, ita si A excludit non-B, ergo includit B, et B excludit non-C, ergo B includit C, itaque denique A includit C.

Si adhibeamus AB est, A non-B est pro particularibus, et A continet B, vel A continet non-B pro universalibus, poterimus carere propositionibus negativis. Sane negativa non afficit copulam nisi quando dicitur propositio esse falsa, alioqui afficit praedicatum.

8th. That a proposition follows from a proposition is nothing other than that the consequent is contained in the antecedent, as a term in a term, and by this method we reduce consequences to propositions, and propositions to terms.

9th. That A contains l is the same as A = xl.[ll]

(199) The *particular affirmative* proposition: AB is.

The *particular negative*: A non-B, is.

And assuming that A and B are, the *Universal Affirmative* proposition: 'A non-B is not'.

The *Universal Negative*: 'AB is not'.

From this it is immediately evident that there are no more than these propositions, and what are their oppositions and conversions. For P. A. and U. N. are opposed, and the same holds for P. N. and U. A. It is also evident that in the proposition 'AB is' or 'AB is not' each term has the same relation, and that here the simple conversion takes place. We can add: 'non-A non-B is' or 'non-A non-B is not'; but this is in no way different from 'LM is' or 'LM is not', assuming that non-A is L and non-B is M.

The U. A., i.e. 'A non-B is not', is the same as 'A contains B', for that A does not contain B is the same as that A non-B is true. Therefore, that A contains B is the same as 'A non-B is not true'.

⟨(200) If I say 'AB is not', it is the same as if I say 'A contains non-B' or 'B contains non-A', i.e. 'A and B are inconsistent'. Similarly, if I say 'A non-B is not', it is the same as if I say 'A contains non-non-B' or 'A contains B', and analogously 'non-B contains non-A'.

Therefore, in these few propositions the foundations of logical form are contained.⟩

[ll]   Every B is C.   B non-C is not.
    Every A is B.   A non-B is not.
    Every A is C.   A non-C is not.

This consequence from purely negatives, however, although sound, is not evident, unless it is reduced to positive form. From this is clear that this reduction of universal to negative propositions is not so natural. Just as, if A contains B and B contains C, A also contains C, so also, if A excludes non-B, it therefore includes B, and if B excludes non-C, therefore B includes C, and thus, finally, A includes C.

If we employ 'AB is', 'A non-B is' for particular propositions, and 'A contains B' or 'A contains non-B' for universal propositions, we shall be able to avoid the use of negative propositions. The negative proposition, indeed, does not modify the copula, except when we say that the proposition is false; otherwise it modifies the predicate.

Commentary

Let us omit for the present...like the _Rainbow_: For the notion of 'I' as denoting a substance, cf. GP2: 52 (Leibniz's letter to Arnauld). _Concrete terms_ of the (English) language are substantives and adjectives like 'man', 'woman', 'tree', 'good', 'slim', etc., whereas the _abstract terms_ of the philosophical tradition of which Leibniz is thinking in this case are terms like 'manhood', 'womanhood', 'goodness', etc. As a 'moderate' nominalist, Leibniz believes that it would be better to abstain from employing abstract terms in philosophy, even though on some occasions he softens his intransigence, admitting that it is very difficult to avoid abstract terms completely (cf. A VI, 4A: 196). On Leibniz's nominalism, see Burkhardt (1980: 410–14); Mates (1986: 170–83); Mugnai (1990). On abstract terms, see further, pp. 50, 112–13.

the being B of A: 'τό' is the neuter singular of the Greek definite article, corresponding to English 'the'. Leibniz usually employs it as the equivalent of quotation marks to distinguish, for instance, the mention of a word from its use. With the same purpose, medieval logicians employed the French definite article 'ly'.

Non-A is _Privative_: Cf. A VI, 4A: 405: 'The _Privative_ is that which indicates a negation'; and _NE_: 130: 'But I should still think that the _idea_ of rest is privative, that is, that it consists only in negation. It is true that the act of denial is something positive.'

'being': The notion of _being_ belongs to the simplest terms from which all others derive. It is equivalent to something positive and possible, because it can be thought without implying a contradiction (A VI, 4A: 388, 400, and 558).

the nothing has no properties: Cf. A VI, 4A: 146, 625, 875, and 551: 'Suppose that _N_ is not _A_, _N_ is not _B_, _N_ is not _C_, and so on. Then we can say that _N_ is _Nothing_. This is what the common saying, that what is not has no attributes, refers to.' On this point, see Mates (1986: 98–100); Angelelli (2006).

Thus, we do not need...for indicating emphasis: Leibniz observes that both substantives and adjectives presuppose some _being_ of which they are predicated. Thus 'man' properly means 'a being which is a man', and 'good' refers to a being which is good (in this respect, there is no difference between substantive and

136 COMMENTARY TO PAGES 43–47

adjective). The 'completeness' that Leibniz mentions on this occasion is not the same as the completeness of the 'complete concept' as defined in the *Discourse on Metaphysics*, which constitutes one of the central notions of Leibniz's metaphysics (see Introduction).

Thus, 'man' is a being in itself... accidental beings: Leibniz disregards here the distinction between *use* and *mention*: to be necessary or changeable are the beings denoted by the terms, not the terms.

And what has to be added... the integrity of the term: According to the ancient Latin grammarians, an *oblique term* is a term in a case other than the nominative. Thus, *homo* ('man'), for instance, is a *direct term*, whereas *hominis* ('of the man') and *homini* ('to the man') are *oblique terms*. Therefore, the phrase 'of Evander' (*Evandri*), when joined with the word 'sword' (*ensis*), is added *obliquely*.

characteristic: 'Characteristic' is the name Leibniz uses to designate the 'symbolic language' of his logical calculus. See Introduction, p. 3.

The progress of this work... may be carried out: Leibniz, however, did not develop this part of the essay.

Therefore, ... partial terms and particles: particles are mainly conjunctions and prepositions.

With this method... characteristic number: On the project of using numbers to represent concepts, see Introduction, p. 2, footnote 4.

(2nd) Simple particles, **i.e. primitive syncategorematic terms:** As we read in Ockham's *Logic*, which repeats on this point the prevailing doctrine among medieval logicians, a *categorematic term* is a term with a *definite* and autonomous meaning, like 'man', 'dog', or 'white', which mean, respectively, men, dogs, and white things. A *syncategorematic term*, instead, is a term that, to have a meaning, needs to be added to a categorematic one. This is the case, for instance, of terms like 'every' and 'some', which, once associated with a categorematic term such as 'man', give rise to phrases like 'every man' and 'some men', which mean respectively every man and some men (see Ockham 1974: 15).

(10th) **We may similarly... which are related to which:** Leibniz is aware that particles involve relations; therefore, if we omit particles, it becomes difficult to distinguish an order between the words in a sentence.

'Term' : In a short essay devoted to the universal calculus written around 1679, Leibniz gives the following explanation of what he means by the word 'term': 'By "term" I understand, not a name, but a concept, i.e. that which is signified by a name; you could also call it a notion, an idea' (*LP*: 39; A VI 4A: 288).

'*T*': We find analogous remarks concerning the notion of 'myself' (*moi*) in a letter to Arnauld belonging to the same period of the *GI*; see GP 2: 52-3.

But which of these solutions... as we progress: Contrary to what Leibniz writes, in what follows he never develops this issue. The entire paragraph summarizes what Leibniz says in the *Meditations on Knowledge, Truth, and Ideas*, published in 1684 on the *Acta Eruditorum* (A VI, 4A: 585-92; L: 291-5).

In particular... no further resolution: Of this opinion were Descartes and his followers. For Leibniz's criticism of Descartes' position, see GP 6: 584. For a discussion of Leibniz's thesis on the reducibility of the notion of *extension*, Parkinson (1965: 166); Broad (1975: 55); Adams (1994: 232-4); Hartz (2007: 63-6).

Hence it follows... coincides with A: Leibniz distinguishes two kinds of identities: (a) a *formal identity*, when the same terms appear on both sides of the *copula* or of the equals sign, as in 'A is A' or 'AB = AB'; (b) a *virtual identity*, or *coincidence*, when the terms involved are not formally the same, as, for instance, in 'A is B' or 'AB = C', but they can be reduced to a formal identity by means of analysis.

Therefore, if B = non-A... some A is B: In the manuscript, Leibniz first wrote 'different' (*diversa*), then deleted it and wrote 'disparate' (*disparata*). Usually Leibniz defines as *different* those terms that are not the same (A, VI, 4A: 294). In the *Addenda to the 'Specimen of the Universal Calculus'* (A VI, 4A: 294), he writes that '*a* and *b* are *disparate* if *a* is not *b* and *b* is not *a*, such as *man* and *stone*' (*LP*: 44). Cf. also the *Elements of Calculus*: 'If neither term is contained in the other, they are called *disparate*' (*LP*: 21; AVI, 4A: 200). On disparate terms in relation to judgement, see Schneider (1980).

I say that something is impossible... incomplex term: Leibniz calls 'incomplex' (i.e. 'not complex') a term in the proper sense of the word (such as 'man' or 'dog'), and 'complex' a term that corresponds to a proposition.

If A = something true... contains non AY: Leibniz's use of drawing a line above a group of letters or words is equivalent to putting them in brackets. On 'indefinite letters' like 'Y', see Introduction.

These four axioms: That is, principles 6-9.

(16)... as Aristotle says: For analogous references to Aristotle, see *DAC* (A VI, 1: 183) and *NE*: 487.

(19) If A is B... a distinction between them: Leibniz calls 'reflexive' or 'indirect' (LH XXXV 1, 14, Bl. 7v; Mugnai 1992: 147) those contexts that in contemporary logic and philosophy of logic are known as 'intensional contexts', where the principle of substitutivity of equivalents fails. See Introduction, pp. 18-9.

(20) Something...lest confusion arise: This is a fundamental clause about the use of variables in a logical calculus. To understand why it is so important, assume: (1) A is B, and (2) C is B. Then, in virtue of §16 we may have: (1') A = BY and (2') C = BY; and from (1') and (2') we may think of deriving (3') A = C. But suppose A = *man*, C = *dog*, and B = *animal*. In this case, the *indefinite* letter 'Y' has an ambiguous meaning, because in (1') it designates the concept that, added to the concept of *animal* gives rise to the concept of *man*, whereas in (2') it designates the concept that, added to the concept of *animal*, produces the concept of *dog*: obviously, the two concepts are different and the conclusion is wrong.

(25) That A is B...some B are A (by 17): Leibniz here attempts to prove the classical rule of the so-called *conversion by accident* (*conversio per accidens*): from 'Every A is B' to infer 'Some B are A'. The proof, however, is faulty: oddly enough, Leibniz contravenes the prescription he himself stated at §23 concerning the use of the indefinite letter 'Y'.

(28) A term set forth by itself: I accept here O'Briant's translation (O'Briant 1968: 41).

(29) 'A is B', therefore 'some A is B': Leibniz here attempts to prove the so-called rule of *subalternation*, i.e. the inference from 'Every A is B' to 'Some A are B'. Usually, for this inference to be valid it needs the so-called 'existential import': it requires, in other words, that the concept of the subject of the universal premise of the inference should be not empty. Leibniz, however, adopting the *intensional* point of view, does not presuppose the existence of the individuals he quantifies on (see Introduction, p. 15).

(55) If A contains B and A is true, then B is also true: This corresponds to the classical rule of the *modus ponens* in propositional logic: 'If *p*, then *q*; *p*; therefore *q*.'

I consider the whole syllogism as a proposition also: According to the prevailing scholastic doctrine, a syllogism was an argument composed of three distinct propositions: two premises and a conclusion. Leibniz here, saying that he considers 'the whole syllogism as a proposition' is probably thinking of it as the conditional which has as antecedent the conjunction of the two premises and the conclusion as consequent.

(56)...a long continuation of the resolution: This is the first embryonic form of Leibniz's solution to the problem of contingency, based on the analogy with the calculus (see Introduction, pp. 36–40). Even if the analysis of a contingent proposition does not reach the first terms and can be carried on without end,

Leibniz thinks that we may find a general rule of the progression which allows us to dispense with a calculation that is impossible to be performed.

(57)...And so in order to ascertain that something is false...it cannot be demonstrated that it is true: Leibniz's remarks on this point are quite tentative: if we cannot prove that a given proposition *p* is true, it does not follow that we have proved that *p* is false.

(72) Therefore, suppose that we have BY: In the *Discourse on Metaphysics* Leibniz introduces the notion of a complete concept as follows: 'we can say it is the nature of an individual substance or complete being to have a concept so complete that it is sufficient to make us understand and deduce from it all the predicates of the subject to which the concept is attributed' (L: 307).

(80) We must see...infinite terms: An *infinite term* is a term immediately preceded by 'non', as 'non-man', 'non-animal', etc. It was called 'infinite' because of the infinite multiplicity of things the denied term refers to: *non-man*, for instance, refers to everything that is not a man.

(82) Of course, it will also be possible...'B = Y non-A': This inference is valid only if B denotes an individual or a complete concept; it is invalid, on the other hand, if B denotes a general concept such as that corresponding to 'man': if 'every man is white' ('man is white') is false, from this it does not follow that every man is non-white. Leibniz corrects this mistake at §92. Cf. Lenzen (2004a: 133–79).

(87)...infinite affirmative and negative propositions: 'Infinite propositions' were considered those propositions with infinite terms as subject or predicate.

(89) Let us consider...successfully performed: To sum up, from §83 to this point, Leibniz represents as follows the four categorical propositions of classical syllogistic:

(A) $A = AB$ (E) $AY \neq ABY$
(I) $AY = ABY$ (O) $A \neq AB$.

(92) The consequence: 'if A is not non-B, then A is B' is invalid...every animal is a man: At §82 above, Leibniz stated the equivalence 'A is not B = A is non-B', which consists of two halves:

$$(82.1)\,(A \text{ is not B}) \text{ is} (A \text{ is non-B})$$

$$(82.2)\,(A \text{ is non-B}) \text{ is} (A \text{ is not B}).$$

Clearly, (82.2) is right, whereas (82.1) is wrong (if A is not included in B, it does not follow that A is included in the logical complement of B). Now, Leibniz here claims that (92.1) '(A is not non-B) implies (A is B)' is false. But (92.1) is the contrapositive of (82.1), which is logically equivalent to this. Therefore, Leibniz implicitly recognizes that (82.1) is wrong. On this point, see Lenzen (2004a: 133–79).

(93) If A is B, non-B is non-A: Leibniz assumes here that 'non-B is non-A' is equivalent to 'non-B is not A', repeating the mistake of §82 and contravening what he had stated immediately above in §92.

(94) If non-B is non-A: Again an instance of (82.1).

(101) If it is false that some A is B, it is false that 'every A is B': Implicit here is: 'is true'.

(107): Leibniz now attempts to elaborate a system of symbols for encoding information about the syntactic structure of terms and propositions involved in the calculus. The symbols employed are numbers and lines; the lines are written over the letters representing terms and propositions. Note that Leibniz uses the form of a tree to represent the structure of a complex proposition, with the simplest expressions at the roots.

Every complex of propositions... the mode of employing the term [C]: The lines drawn over the letters may be considered as playing the role of brackets: thus, the proposition of the diagram can be represented as '((AB)C)'. The diagram shows that the proposition has been composed making two steps: first putting together A and B, then adding C to the complex (AB).

⟨**'A is this true thing'**⟩: To understand the analogy between *truth* and *unity* established by Leibniz, we need to see how the diagram with the Sun can be transformed into the diagram with the Moon. Leibniz assumes that if a complex proposition is composed of two or more propositions, each composing proposition should be decomposed into the same number of elements as the others. Thus, suppose we have four terms 'A', 'B', 'C', and 'D' and that the complex 'AB' corresponds to a proposition. Adding C to AB, we give rise to a second proposition, which, using brackets instead of lines, may be represented as '((AB)C)'. This second proposition is composed of two parts of different 'weight': AB and C; to equalize the two parts, Leibniz introduces the symbol 'V', corresponding to 'true', and adds it to C: ((AB)(CV)). If we now want to add the term 'D' to this last expression, keeping close to the principle of the same number of parts, we should add 'DV' to '((AB)(CV))', thus obtaining '(((AB)(CV))(DV))'. This expression, however, is still 'unbalanced', because '((AB)(CV))' is made up of two

propositions, whereas '(DV)' is made up of two terms, i.e. it corresponds to a proposition only. Therefore, to 'balance' the complex proposition again, we need to add the couple 'VV' to the previous expression, thus obtaining: '(((AB)(CV)) ((DV)(VV)))'. Now, it is easy to see that the diagram with the symbol of the Moon is obtained from that with the symbol of the Sun integrating into it with a suitable number of 'V's any expression that cannot be decomposed according to the principle of the same number of parts. Quite interesting is Leibniz's remark that 'true' plays here the same role of unity in arithmetic: as in a multiplication the number of factors can be increased by joining several unities to them (for instance: $4 \times 2 \times 1 \times 1 = 4 \times 2$); thus, nothing changes if we add 'true' to a term or to a proposition.

(110) **New reflexive terms . . . by some name:** Leibniz here calls 'reflexive' those terms that refer to other terms employed in the language of the logical calculus.

(112) **. . . for it cannot be anything else:** At §81 above, Leibniz defined 'Y' as 'one indeterminate thing' and '$\overline{Y}$' as 'anything'. Here he interprets 'some A is B' as meaning '*this* particular A is B', i.e. as if 'some' should refer to a particular item of the domain of A, and raises the question of the relationship that '*this* particular Y' has with 'any Y'. Then, he argues that 'any Y' contains '*this* Y' (that is: $\overline{Y}$ contains Y). According to the Aristotelian square, however, the negation of the particular affirmative proposition (I: 'some A is B') implies the universal negative (E: 'no A is B'): but if the particular affirmative is expressed as 'AY is B' and the universal negative as 'A$\overline{Y}$ is not B', this implication does not hold, whereas it holds in the opposite direction: 'not (A$\overline{Y}$ is not B)' implies 'AY is B'—in clear contrast to the inferences of the traditional Aristotelian square.

(113): The dotted part of line B means that B could coincide with A. For a very useful and clear exposition of Leibniz's various attempts to represent inferences by means of diagrams, see Lenzen's introductory remarks to the selection of texts in *Schriften zur Syllogistik*: 332–453.

⟨A small perpendicular line . . . lines drawn still farther down: Leibniz wrote these remarks in the margin of the manuscript, beside §§114–21: they partly complete the text (see, in particular, the explanation concerning the meaning of the lines) and partly sound like a private comment made with a view to a revision.

(120): Leibniz has drawn this diagram from the *extensional* point of view (see Introduction, pp. 13–5.), whereas he has built the preceding diagrams from the *intensional* one. If we deny that some A is B (the particular affirmative, I), we are stating at the same time that it is true that no A is B (universal negative, E, the contradictory of I), a situation that is well represented by the diagram,

according to which the two lines A and B do not share a single point. The extensional point of view, indeed, considers the individuals (the points, in this case) denoted by the terms (A and B), whereas the intensional point of view considers the conceptual parts composing a given concept. Thus, the proposition 'No man is a dog', for instance, can be shown to be true from the extensional point of view, because no individual belongs to both the set of men (associated with the concept of man) and the set of dogs. If we consider, on the other hand, the same proposition from the intensional point of view, the diagram cannot represent it, because the concept of man and that of dog have several conceptual parts in common.

Parkinson (*PL*: XL) suggests that here Leibniz may be thinking of the two concepts corresponding, respectively, to A and B as containing conceptual parts that are not compatible (i.e. that give rise to a contradiction). In this case, A and B cannot be put together, and the diagram depicts quite fairly the circumstance that A excludes B. In a later essay entirely devoted to the representation of logical inferences by means of diagrams (*De formae logicae comprobatione per linearum ductus* [*On the proof of logical form by the drawing of lines*] 1702?), however, Leibniz proposes a different solution to the problem of expressing the categorical universal negative proposition from the intensional point of view. As Leibniz remarks in §16 above, any categorical universal affirmative proposition of the general form '(Every) A is B' can be expressed as an equivalence: A = BX, where 'X' represents some conceptual content that joined to B produces a concept equal to A. Thus, for instance, we may transform '(Every) man is an animal' into 'man = rational animal'. In the essay on diagrams, Leibniz transforms the universal negative proposition into a universal *affirmative* proposition with a denied predicate. Thus 'No man is a stone' becomes '(Every) Man is a non-stone', and this is the corresponding diagram as built by Leibniz:

$$\text{Man} = \text{non-stone}............\text{X}$$
$$\text{non-stone}$$

In other words, Leibniz assumes that the negative concept of being a non-stone inheres in the concept of man.

(122)... **the individuals of the genus**: Here Leibniz explicitly adopts the extensional point of view.

(123)... **And this representation is the inverse of the preceding one**: That is the representation based on the intensional point of view.

(124)... **by means of numbers**: Leibniz thought for the first time of employing numbers to denote concepts in his 1666 *Dissertation on Combinatorial Art* (see

Introduction, p. 2, footnote 4). At §§124–9 and 187 he comes back to this old project without fully developing it.

A, A², A³ are considered as the same: Leibniz stresses here that the operation of juxtaposition of letters obeys the principle of *idempotence* (see Introduction, p. 22).

(125)...cannot be divided by B: Leibniz represents as '1/B' the negation of B. Thus, if A = C/B and C cannot be divided by B, by putting together B and 1/B a contradiction arises: in this case Leibniz says that B is 'destroyed' in AB.

(127)...be multiplied by B: Assuming A as Subject and B as predicate, according to what Leibniz says in §§124–7, we have:

(1) (A) – '(Every) A is B' is true if the number corresponding to B divides the number corresponding to A (§124).

(2) (I) – 'AB is B' ('Some A is B') is true if B divides AB (§125).

(3) (E) – 'No A is B' is true if it is false that B divides AB (§126).

(4) (O) – 'Some A is not B' is true if '(Every) A is B' is false, i.e. if it is false that B divides A (§127).

(129)...the particular affirmative arises: That is:

$$(A)\ A:B \qquad\qquad (E)\ A:non\text{-}B$$

$$(I)\ not(A:non\text{-}B) \qquad\qquad (O)\ not(A:B).$$

Thus, I have discovered that secret...several years ago: What exactly Leibniz is referring to here is difficult to say: probably he has in mind some essay on logical calculus of the period 1679–80.

(131)...whose complete demonstration transcends every finite intellect: Here Leibniz seems implicitly to admit that God, having an *infinite* intellect, can prove the contingent propositions.

(134)...all contingent truths with certainty: At §131 Leibniz states that God 'knows even the truths of contingent propositions, whose complete proof transcends every finite intellect', and here he specifies that God 'knows all contingent truths with certainty'. In a text entitled *De libertate* (*On liberty*), written some years after the *GI*, Leibniz firmly states that even God *cannot know* the contingent truths by means of a logical proof.

(137)...to explain a little more distinctly this last point: As Parkinson remarks (*LP*: xliv), at §45 Leibniz 'has merely expressed the hope that he will be able to

treat hypothetical propositions as categorical'. Moreover, before §138 there is no trace of any discussion concerning the origin of abstract terms.

(138)... the D-ness of C: On 'logical abstract terms' (i.e. expressions like '*A*'s *being B*', '*C*'s *being D*') and their importance for the logical calculus of the *GI*, see Introduction, pp. 23–6 and Malink and Vasudevan (2016).

(144) Propositions are... i.e. is actually a being: The Latin expression *secundum adiectum* means literally 'second added' and the genitive *secundi adiecti* joined to the word *propositio* ('proposition') gives rise to 'proposition of second added'. Analogously, *tertium adiectum* means 'third added' and *propositio tertii adiecti* means 'proposition of third added'. This is scholastic jargon originating in Aristotle's works and presupposes that the basic structure of a proposition had the form: subject + copula + predicate (cf. Aristotle, *De interpretatione*, 19b19–20; Abelard (1970: 161); Burley (1955: 54)). To be added is the copula—the verb 'to be'—and it is added *as second* if it immediately follows the subject and there is no predicate explicitly expressed in the proposition, as in 'Peter is'. It is added as third if it composes the proposition together with the subject and the predicate (as in 'Peter is a man'). On the genesis of the distinction between *secundi* and *tertii adiecti* (or *adiacentis*), and its development during the scholastic period, see Nuchelmans (1992); for the distinction in Leibniz, see Burkhardt (1980: 132) and also *LP*: xlv, n. 1.

(146)... is essential or existential: Thus, for instance, the proposition *tertii adiecti* 'Socrates is wise' can be transformed into the proposition *secondi adiecti* 'Socrates wise is', in which 'Socrates wise', as Leibniz points out, is a 'thing', i.e. a non-contradictory complex (therefore, something simply possible or even existing).

(153) This, however... a real term: Concerning the *constantia subiecti*, Leibniz writes in the *New Essays*: 'The Scholastics hotly debated de Constantia subjecti, as they put it, i.e. how a proposition about a subject can have a real truth if the subject does not exist. The answer is that its truth is a merely conditional one which says that if the subject ever does exist, it will be found to be thus and so. But it will be further asked what the ground is for this connection, since there is a reality in it which does not mislead. The reply is that it is grounded in the linking together of ideas' (*NE*: 447).

(155)... Therefore AB = ((R)): 'R' is the first letter of the Latin word 'res' (thing) and the double brackets mean that the complex 'AB' is made of two 'things', i.e. of two terms, which are real.

(162)…the following propositions will arise: Leibniz, however, at §165 and from §181 onwards disregards the proposal of omitting the indefinite letter 'Y'.

(171) The principles are: This is only the first set of principles: another two will be proposed, at §189 and §198, respectively.

(189) *Third*, the same…but false: A 'false term', as Leibniz will remark at §194. Analogously, the 'true terms' of §192 are terms which do not imply a contradiction.

(190) The *particular negative*…does not imply opposites: To sum up:

$$(A) \; A = XL \qquad (E) \; A = Y \text{ non-B}$$
$$(I) \; AN = BL \qquad (O) \; AX = Z \text{ non-L.}$$

Bibliography

Abaelardus, Petrus (1970). *Dialectica*, ed. Lambertus Maria De Rijk. Assen: Van Gorcum.

Adams, Robert (1994). *Leibniz: Determinist, Theist, Idealist*. New York and Oxford: Oxford University Press.

Angelelli, Ignacio (2006). 'The Interpretations of "nihili nullae sunt proprietates": A Text from Rubio', in Guido Imaguire and Christina Schneider, eds., *Untersuchungen zur Ontologie*. Munich: Philosophia Verlag, 41–53.

Antognazza, Maria Rosa (2009). *Leibniz: An Intellectual Biography*. Cambridge: Cambridge University Press.

Antognazza, Maria Rosa, ed. (2018). *The Oxford Handbook of Leibniz*. Oxford: Oxford University Press.

Barnes, Jonathan (1983). 'Terms and Sentences.' *Proceedings of the British Academy*, 69: 279–326.

Barnes, Jonathan (2007). *Truth, etc*. Oxford: Oxford University Press.

Bernini, Sergio (2002). 'La logica di Leibniz, i possibili e l'infinito', *Annali del Dipartimento di Filosofia*. Florence: Firenze University Press, 21–51.

Boole, George (1847). *The Mathematical Analysis of Logic*. Cambridge: MacMillan and London: George Bell.

Broad, Charlie D. (1975). *Leibniz: An Introduction*, ed. C. Lewy. Cambridge: Cambridge University Press.

Burkhardt, Hans (1980). *Logik und Semiotik in der Philosophie von Leibniz*. Munich: Philosophia Verlag.

Burleigh, Walter (1955). *De puritate artis logicae. Tractatus longior*, ed. Philotheus Boehner. New York: Franciscan Institute.

Castañeda, Hector-Neri (1974). 'Leibniz's Concepts and Their Coincidence salva veritate', *Noûs*, 8: 381–98.

Castañeda, Hector-Neri (1976). 'Leibniz's Syllogistico-Propositional Calculus', *Notre Dame Journal of Formal Logic*, 17: 481–500.

Castañeda, Hector-Neri (1990). 'Leibniz's Complete Propositional Logic', *Topoi*, 9: 15–28.

Cohen, Jonathan (1954). 'On the Project of a Universal Character', *Mind*, 62: 49–63.

Couturat, Louis (1901). *La Logique de Leibniz d'après des documents inédits*. Paris: Alcan (1961: Hildesheim: Olms).

Cover, J. A., and O'Leary-Hawthorne, John (1999). *Substance and Individuation in Leibniz*. Cambridge: Cambridge University Press.

Di Bella, Stefano (2018). 'The Complete Concept of an Individual Substance', in Maria Rosa Antognazza, ed., *The Oxford Handbook of Leibniz*. Oxford: Oxford University Press, 119–36.

Dummett, Michael (1956). Review of Nicholas Rescher, 'Leibniz's Interpretation of his Logical Calculi', *Journal of Symbolic Logic*, 21: 197–9.

Dürr, Karl (1930). *Neue Beleuchtung einer Theorie von Leibniz*. Darmstadt: Abhandlungen der Leibniz-Gesellschaft, vol. III.

Dürr, Karl (1949). 'Leibniz' Forschungen im Gebiet der Syllogistik', Göttingen: Dieterichsche Universitäts-Buchdruckerei W. Fr. Kaestner.

Dürr, Karl (1955). 'Die Syllogistik des Johannes Hospinianus', *Synthese*, 9: 272–84.

El-Rouayheb, Khaled (2010). *Relational Syllogisms and the History of Arabic Logic, 900–1900*. Leiden and Boston, MA: Brill.

Hailperin, Theodore (1981). 'Boole's Algebra Isn't Boolean Algebra', *Mathematics Magazine*, 54 (4): 172–84.

Hailperin, Theodore (2004). 'Algebraical Logic 1685–1900', in Dov M. Gabbay and John Woods, eds., *The Rise of Modern Logic: From Leibniz to Frege*. Handbook of the History of Logic, Vol. 3. Amsterdam, Boston, MA, and New York: Elsevier North Holland, 323–88.

Hartz, Glenn (2007). *Leibniz's Final System: Monads, Matter and Animals*. London and New York: Routledge.

Jungius, Joachim (1957). *Logica Hamburgensis*, ed. Rudolf W. Meyer. Hamburg: In Aedibus J. J. Augustin.

Kauppi, Raili (1960). *Über die Leibnizsche Logik*. Helsinki: Acta Philosophica Fennica.

Lenzen, Wolfgang (1983). 'Zur extensionalen und "intensionalen" Interpretation der Leibnizschen Logik', *Studia Leibnitiana*, 15 (2): 129–48.

Lenzen Wolfgang (1984a). '"Unbestimmte Begriffe" bei Leibniz', *Studia Leibnitiana*, 16 (1): 1–26.

Lenzen Wolfgang (1984b). 'Leibniz und die Boolesche Algebra', *Studia Leibnitiana*, 16 (2): 188–203.

Lenzen Wolfgang (1986). '"Non est non est non". Zu Leibnizens Theorie der Negation', *Studia Leibnitiana*, 18 (1): 1–37.

Lenzen Wolfgang (1991). 'Leibniz on Ens and Existence', in Wolgang Spohn et al., eds., *Existence and Explanation*. Dordrecht: Kluwer Akademic Publishers, 59–75.

Lenzen Wolfgang (2004a). *Calculus Universalis: Studien zur Logik von G. W. Leibniz*. Paderborn: mentis.

Lenzen Wolfgang (2004b). 'Leibniz's Logic', in Dov M. Gabbay and John Woods, eds., *The Rise of Modern Logic: From Leibniz to Frege*. Handbook of the History of Logic, vol. 3. Amsterdam, Boston, MA, and New York: Elsevier North Holland, 1–83.

McDonough, Jeffrey K. (2018). 'Freedom and Contingency', in Maria Rosa Antognazza, ed., *The Oxford Handbook of Leibniz*. Oxford: Oxford University Press, 86–99.

Malink, Marko, and Vasudevan, Anubav (2016). 'The Logic of Leibniz's Generales inquisitiones de analysi notionum et veritatum', *The Review of Symbolic Logic*, 9: 686–751.

Malink, Marko, and Vasudevan, Anubav (2017). 'Leibniz's Theory of Propositional Terms: A Reply to Massimo Mugnai', *The Leibniz Review*, 27: 139–55.

Malink, Marko, and Vasudevan, Anubav (2019). 'Leibniz in the Logic of Conceptual Containment and Coincidence', in Vincenzo De Risi, ed., *Leibniz and the Structure of Sciences*. Modern Perspectives on the History of Logic, Mathematics, Epistemology. Cham: Springer, 1–46.

Mates, Benson (1972). 'Individuals and Modality in the Philosophy of Leibniz', *Studia Leibnitiana*, 4 (2): 81–118.

Mates, Benson (1986). *The Philosophy of Leibniz. Metaphysics and Language*. New York and Oxford: Oxford University Press.

Mondadori, Fabrizio (1973). 'Reference, Essentialism, and Modality in Leibniz's Metaphysics', *Studia Leibnitiana*, 5 (1): 74–101.

Mondadori, Fabrizio (1975). 'Leibniz and the Doctrine of Inter-World Identity', *Studia Leibnitiana*, 7: 22–57.

Mondadori, Fabrizio (1985). 'Understanding Superessentialism', *Studia Leibnitiana*, 17: 162–90.

Mondadori, Fabrizio (1993). 'On Some Disputed Questions in Leibniz's Metaphysics', *Studia Leibnitiana*, 25: 153–73.

Mugnai, Massimo (1992). *Leibniz's Theory of Relations*, Studia Leibnitiana Supplementa, 28. Stuttgart: Franz Steiner Verlag.

Mugnai, Massimo (2012). 'Leibniz's Theory of Relations: A Last Word?'. *Oxford Studies in Early Modern Philosophy*, 6I: 171–208.

Mugnai, Massimo (2017). 'The Logic of Leibniz's *Generales inquisitiones de analysi notionum et veritatum* by Marko Malink and Anubav Vasudevan'. *The Leibniz Review*, 27: 117–37.

Mugnai, Massimo (2018). 'Ars characteristica, Logical Calculus, and Natural Languages', in Maria Rosa Antognazza, ed., *The Oxford Handbook of Leibniz*. Oxford: Oxford University Press, 177–207.

Nuchelmans, Gabriel (1980). *Late-Scholastic and Humanist Theories of the Proposition*. Amsterdam, Oxford, and New York: Verhandelingen der Koninklijke Nederlandse Akadamie van Wetenshappen.

Nuchelmans, Gabriel (1992). *Secundum/Tertium Adiaciens: Vicissitudes of a Logical Distinction*. Amsterdam: Royal Netherlands Academy of Art and Sciences.

Peckhaus, Volker (1997). *Logik, Mathesis universalis und allgemeine Wissenschaft: Leibniz und die Wiederentdeckung der formalen Logik im 19. Jahrhundert*. Berlin: Akademie Verlag.

Pelletier, Arnaud (2018). 'The Scientia Generalis and the Encyclopaedia', in Maria Rosa Antognazza, ed., *The Oxford Handbook of Leibniz*. Oxford: Oxford University Press, 162–76.

Rescher, Nicholas (1954). 'Leibniz's Interpretation of his Logical Calculi', *Journal of Symbolic Logic*, 19: 1–13.

Sauer, Heinrich (1946). 'Über die logischen Forschungen von Leibniz', in G. W. Leibniz, *Vorträge der aus Anlass seines seines 300. Geburtstages in Hamburg abgehaltenen wissenschaftlichen Tagung*. Hamburg: Hansischer Gildenverlag, Joachim Heitmann & Co., 46–78.

Schneider, Martin (1970). *Analysis und Synthesis bei Leibniz*. Bonn: Rheinische Friedrich-Wilhelms-Universität.

Schneider, Martin (1980). 'Das Problem des Disparaten Sätzen in Leibniz's Logik', *Studia Leibnitiana Supplementa*, 13: 23–32.

Schröter, Karl (1973). 'Die Beiträge von Leibniz zur Algebra der verbandstheoretischen Relationen und Operationen', *Studia Leibnitiana Supplementa*, 2: 27–36.

Schupp, Franz (1993). *Introduction to G. W. Leibniz, Die Grundlagen des logischen Kalküls*. Hamburg: Meiner Verlag.

Sleigh Jr, Robert C. (1990). *Leibniz and Arnauld: A Commentary on their Correspondence*. New Haven, CT: Yale University Press.

Sommers, Fred (1982). *The Logic of Natural Language*. Oxford: Clarendon Press.

Sommers, Fred (1993). 'The World, the Facts, and Primary Logic', *Notre Dame Journal of Formal Logic*, 34: 169–82.

Sotirov, Vladimir (1999). 'Arithmetizations of Syllogistic à la Leibniz', *Journal of Applied Non-Classical Logics*, 9: 387–405.

Swoyer, Chris (1995). 'Leibniz on Intension and Extension', *Noûs*, 29: 96–114.

Thom, Paul (1981). *The Syllogism*. Munich: Philosophia Verlag.

Vagetius, Johannes (1977). *Logicae Hamburgensis Additamenta*, Göttingen: Vandenhoeck & Ruprecht.

William of Ockham (1974). *Summa logicae*, ed. P. Boehner, G. Gal, and S. Brown. New York: St. Bonaventure.

Index Nominum

General Index